ROAD MAP TO SCALE UP SOLAR IRRIGATION PUMPS IN BANGLADESH (2023–2031)

DECEMBER 2023

ADB

Contents

Tables and Figures

Figures

Foreword

The Asian Development Bank (ADB) has been supporting Bangladesh's development since 1973 with cumulative commitments of $33.66 billion as of the end of December 2022. As a key source of external assistance for Bangladesh, ADB has been providing an average of $2 billion per year since 2016, aligning its assistance with the country's Eighth Five-Year Plan, 2021–2025, and the Perspective Plan, 2021–2041. In 2022, ADB committed $1.6 billion in public sector loans and grants to Bangladesh, including $250 million for social protection and $200 million in microenterprise financing for job creation to support recovery from the coronavirus disease (COVID-19) pandemic.

Bangladesh aims to scale up solar irrigation pumps (SIPs) to boost agricultural productivity and reduce reliance on fossil fuels. Recognizing the energy, water, and food nexus, ADB's assistance aims to help Bangladesh maintain steady and robust economic growth.

This Road Map to Scale Up Solar Irrigation Pumps (2023–2031) outlines strategic actions, policy interventions, and stakeholder collaborations required for a successful transition and sets the stage for a greener and more sustainable future.

The road map aims to guide Bangladesh in adopting SIPs to improve its agriculture sector performance and address water and energy challenges. It incorporates inputs from diverse experts and offers strategies to enhance rural livelihoods, reduce emissions, and increase climate change resilience. Replacing diesel pumps with SIPs in Bangladesh can displace the consumption of 1 million tons of diesel fuel annually and reduce 3 million metric tons of carbon dioxide equivalent every year. This transition offers economic and environmental benefits. Adopting the road map would support Bangladesh in achieving a sustainable and secure water and energy supply for irrigation and providing a new steady income source to farmers.

The SIP road map includes plans for policy frameworks, financing, infrastructure development, capacity building, stakeholder engagement, and monitoring. Collaboration and partnerships are crucial for its successful implementation.

I want to thank everyone who has dedicated their time and expertise in creating this road map. It serves as a guide and a call to action for promoting a greener and more prosperous Bangladesh. Let's work together toward a resilient and inclusive future, where solar energy and sustainable agriculture play a key role.

Kenichi Yokoyama
Director General
South Asia Department
Asian Development Bank

Acknowledgments

The road map technical study was prepared by a team led by Hongwei Zhang, Principal Portfolio Management Specialist/Mission Leader, Energy Sector Office, Sectors Group (SG-ENE), and Marie L'hostis, Water Resources Specialist/Co-Mission Leader, Agriculture, Food, Nature, and Rural Development Sector Office, Sectors Group (SG-AFNR). The technical team comprises Olivier L. Drieu, Senior Water Resources Specialist, SG-AFNR; Geoffrey Wilson, Senior Water Resources Specialist, SG-AFNR; Nazmun Nahar, Senior Project Officer (Energy), Bangladesh Resident Mission, South Asia Department; Maria Charina Apolo Santos, project analyst, SG-ENE; and a group of energy experts (Anthony J. Jude, Sergio Ugarte, Liu Zuming, and Mahmud Sohel) and water and agriculture experts (Prathapar Sanmugam, Trevor Beaumont, Asad Zaman, and Md. Mushfiqur Rahman). The team is grateful for the valuable guidance and support of Kenichi Yokoyama, Director General, South Asia Department; Priyantha Wijayatunga, Senior Director, SG-ENE; Edimon Ginting, Country Director, Bangladesh Resident Mission, South Asia Department; Sujata Gupta, Director, SG-ENE; Mio Oka, Director, SG-AFNR; and Neeta Pokhrel, Director, Water and Urban Development Sector Office, Sectors Group. Financial support is acknowledged with great appreciation from the People's Republic of China Poverty Reduction and Regional Cooperation Fund and the Netherlands Trust Fund under the Water Financing Partnership Facility, and the Netherlands.

The team is pleased to note that the government approved this Road Map to Scale Up Solar Irrigation Pumps in Bangladesh (2023–2031) and published it on 28 November 2023 on their website as well.

The team is grateful for the guidance and collaboration of the Power Division, the Ministry of Power, Energy and Mineral Resources, the Ministry of Agriculture, and the Ministry of Water Resources, Bangladesh. Comments received from the Sustainable and Renewable Energy Development Authority, Bangladesh Agriculture Development Corporation, Barind Multipurpose Development Authority, Bangladesh Water Development Board, Department of Agriculture Extension, Infrastructure Development Company Limited, Local Government Engineering Department, Rural Development Academy, and other private stakeholders were very valuable in formulating the road map for implementation.

Abbreviations

AC	alternating current
ADB	Asian Development Bank
BADC	Bangladesh Agriculture Development Corporation
BERC	Bangladesh Energy Regulatory Commission
BMDA	Barind Multipurpose Development Authority
BREB	Bangladesh Rural Electrification Board
BWDB	Bangladesh Water Development Board
BWM	Bengal Water Machine
CWR	crop water requirements
DAE	Department of Agriculture Extension
DC	direct current
DTW	deep tube well
GCF	Green Climate Fund
GDP	gross domestic product
GHG	greenhouse gas
GIS	Geographical Information System
GWS	groundwater storage
IDCOL	Infrastructure Development Company Limited
LGED	Local Government Engineering Department
LLP	low-lift pump
MCA	multi-criteria analysis
MOPEMR	Ministry of Power, Energy and Mineral Resources
MPPT	maximum power point tracking
NDC	nationally determined contribution
O&M	operation and maintenance
PV	photovoltaic
RDA	Rural Development Academy
SDG	Sustainable Development Goal
SIP	solar irrigation pump
SREDA	Sustainable Renewable Energy Development Authority
STW	shallow tube well
TWS	terrestrial water storage
VRE	variable renewable energy
WMO	water management organization

WEIGHTS AND MEASURES

GWh	gigawatt-hour
GWp	gigawatt peak
ha	hectare
kg	kilogram
kV	kilovolt
kW	kilowatt
kWh	kilowatt-hour
kWp	kilowatt peak
L	liter
mm	millimeters
$MtCO_2e$	metric tons of carbon dioxide equivalent
MW	megawatt
MWp	megawatt peak
Wp	watt peak

In this publication, "$" refers to United States dollars.

Currency Equivalent
$1.00 = 107.73 Tk
(9 July 2023)

Executive Summary

Introduction

Irrigation costs in Bangladesh account for 43% of total agricultural costs. Minor irrigation is done with 1.22 million diesel pumps and more than 430,000 electric pumps. Replacing diesel pumps with modern and efficient solar irrigation pump (SIP) systems is an appropriate and sustainable solution that reduces irrigators' dependency on imported diesel fuel and enables a transition to clean energy. Introducing SIP systems will enable Bangladesh to gradually displace the consumption of 1 million tons of diesel fuel annually, thereby avoiding 3 million metric tons of carbon dioxide equivalent ($MtCO_2e$) every year. Several solar irrigation pilot interventions have occurred in Bangladesh in recent years. Lessons learned in these pilot interventions include the following: (i) the importance of adequately identifying targeted beneficiaries; (ii) the importance of good site selection and system sizing; (iii) high investment costs and low uptake will lead to requiring subsidies; (iv) high operation costs and low utilization plans occur after the Boro rice season; (v) sufficient provisions must be made for operation and maintenance (O&M); (vi) farmers have a reduced interest in keeping SIP systems after receiving grid connection; and (vii) there is a risk of theft and vandalization of equipment.

The Asian Development Bank (ADB) has provided $42.4 million to support the Government of Bangladesh's efforts in converting existing diesel pumps to solar photovoltaic irrigation pumps. Of the amount provided, $20.244 million is a grant from the Scaling-Up of Renewable Energy Program. This program suggested a structured long-term approach, given the government's ambition to replace diesel pumps with SIPs as a condition for their support. In this context, the Sustainable and Renewable Energy Development Authority (SREDA) should be given a project coordination role to align efforts of all different implementing agencies, such as Bangladesh Agriculture Development Corporation (BADC), Barind Multipurpose Development Authority (BMDA), Bangladesh Water Development Board (BWDB), Department of Agriculture Extension (DAE), six electricity distributing utilities,[1] Infrastructure Development Company Limited (IDCOL) Local Government Engineering Department (LGED), Rural Development Academy (RDA), and other agencies implementing or financing SIP systems with public funding to establish guidelines to scale up the use of SIPs. This proposed road map has been developed in coordination with several government agencies.[2]

---

[1] Bangladesh Power Development Board (BPDB), Bangladesh Rural Electrification Board (BREB), Dhaka Electricity Supply Company (DESCO), Dhaka Power Distribution Company (DPDC), Northern Electricity Supply Company Limited (NESCO), and West Zone Power Distribution Company Limited (WZPDCL). The six electricity distributing utilities will be responsible for ensuring grid integration of future solar pumps.

[2] These agencies include SREDA; Power Division of the Ministry of Power, Energy and Mineral Resources; BADC; Bangladesh Agriculture Research Institute; BMDA; Bangladesh Rural Electrification Board; Infrastructure Development Company Limited; the DAE; and the RDA.

Scope

This road map aims to provide guidelines for installing up to 45,000 SIP systems with pump ratings ranging from 4 kilowatts (kW) to 25 kW that would add up to 1,000 megawatt peak (MWp) of solar capacity to the country. These SIP systems would displace the consumption of 300,000 tons of diesel fuel annually, avoiding 900,000 metric tons of carbon dioxide equivalent ($MtCO_2e$) emissions every year. The target number of SIP systems aims to replace diesel pumps irrigating up to 400,000 hectares of land and serving more than 1.3 million farmers. The indicative breakdown per type of SIP system proposed is 15,000 solar low-lift pumps for irrigation with surface water, 2,000 solar deep tube wells, and 28,000 shallow tube wells for groundwater irrigation. The proposed number of solar pumps will produce surplus electricity of 480 gigawatt-hours (GWh) per year for export to the grid, almost 1% of Bangladesh's current total electricity generation. This would provide a regular income to farmers. Studies conducted for the Power Division of the Ministry of Power, Energy and Mineral Resources in 2021 concluded that load flow, voltage stability, frequency response, and angular stability for 10% and 25% integration of variable renewable energy (VRE), such as solar and wind energy, do not negatively affect the grid, so the grid would remain stable. Since this road map proposes the integration of much lower shares of VRE, the conclusion is that the road map's rollout does not represent any significant risk to the stability of the grid.

The SIP target proposed by the road map is significantly more ambitious than interventions foreseen in the agriculture sector by Bangladesh's Nationally Determined Contribution (NDC) and may require development partner support from the Asian Development Bank (ADB), World Bank, Agence Française de Développement, Deutsche Gesellschaft für Internationale Zusammenarbeit, Green Climate Fund, International Solar Alliance, Islamic Development Bank, and Japan to raise the estimated financing of $800 million in a mix of loans and grants.

Irrigation Sources

Cropping intensity is gradually increasing in Bangladesh, putting pressure on irrigation requirements. While Bangladesh has abundant surface water (rivers, lakes, and ponds), rivers and canals often dry up in March and April and at this time surface water becomes scarce. The development of irrigation with surface water is constrained due to a lack of storage possibilities. The purpose of the dams, barrages, and canals built by the Bangladesh Water Development Board (BWDB) is to lessen the destructive impact of floods and utilize excess water for irrigation.

Suitable Areas for Solar Irrigation Pump Rollout

The road map recommends SIP rollout for surface irrigation only in *upazilas* (subdistricts) where surface water availability remains perennial. These include major irrigation projects in *upazilas* where rubber dams are in place. *Upazilas* where temporary storages are present (village ponds) are excluded. The road map identifies 75 out of 495 *upazilas* as high-potential locations for surface water irrigation in which the existing electrified low-lift pumps can be replaced with SIP systems. Through a geographic information system (GIS)-based multi-criteria analysis looking at seven biophysical factors based on available public data, the road map also identifies 350 out of 495 *upazilas* that are suitable for groundwater irrigation. These biophysical factors include salinity, flood depth, tidal surge, tube well spacing, groundwater storage change, and Boro crop water requirement. Out of the 350 *upazilas* suitable for groundwater irrigation, the current recharge mechanisms are adequate to sustain an SIP rollout in 158 of them (locations with potential, level 1). The analysis also concludes that water conservation measures are needed to support groundwater irrigation in another 137 *upazilas* (locations with potential, level 2), and another 55 *upazilas* are suitable with caution and only after a detailed analysis of all seven selection factors at each proposed site is carried out (locations with potential, level 3). Finally, the road map has identified 145 *upazilas* where the rollout of SIP systems using groundwater is unsuitable.

Working Models

Agencies promoting SIPs have experimented with different financing-working models, which can be categorized into the following three models, based on ownership and water source:

(i)   Community/agency-based surface water irrigation, used by the BADC and BMDA
(ii)   Community/agency-based groundwater irrigation, used by the BADC, IDCOL, nongovernment organizations, and private companies
(iii) Individually owned groundwater irrigation, used by the Bangladesh Rural Electrification Board (BREB) and other utilities

The general economic principle the road map rollout follows is that the SIP-based irrigation cost of targeted installations should not exceed the farmers' existing diesel-operated pumps' irrigation cost.

Under the community-based groundwater irrigation model, farmers' groups can reduce the number of tube wells and share project costs. Two framework modalities exist under this model: (i) direct ownership by farmers and (ii) fee-for-service by a private project sponsor, usually for a profit, but including O&M for the SIPs.

In the direct ownership modality, farmers organize through an irrigation committee or village cooperative, which owns and fully manages the operation of the SIP systems. In the fee-for-service modality, it is an intermediary or project sponsor who owns and manages the SIP systems for a profit. The sponsor sells the irrigation water to farmers nearby and may use the surplus power generated in commercial activities to increase the agricultural productivity of their clients. The fee-for-service modality suits well the larger system sizes targeting medium-sized farms that complement their subsistence farming activities with agribusiness activities.

The individually owned groundwater irrigation model requires solid post-rollout technical support, as learned from BADC, BREB, and LGED pilots. Under this model, the owner may ask their neighbors to share irrigation services, a common practice in South Asia.

Regardless of the financing-working model, a minimum 50% grant of the benchmark cost will be required to entice farmers to make the switch to SIP systems. However, government agencies like the BADC, BMDA, and DAE have indicated that they need about 65% in grant financing to entice farmers to make the switch from diesel pumps to SIP systems. The road map suggests that to implement SIPs, the government should (i) decide on a suitable grant financing level (e.g., 50%, 65%, or higher) and budget for it and (ii) establish mechanisms (like insurance) to help farmers repay loans when they are in financial trouble or following disasters like flooding and cyclones that may damage crops and SIP systems.

The possibility of exporting surplus electricity to the grid exists and will provide a regular income to farmers and project sponsors. This road map recommends establishing power purchase agreements between distribution utilities and farmers with SIP systems that initially will only export electricity to the grid and not consume electricity from the grid. For this case, distribution utilities pay for the exported electricity at the 33 kilovolt bulk electricity tariffs. The road map recommends that the Bangladesh Energy Regulatory Commission (BERC) assesses the convenience of higher electricity tariffs than the current 33 kilovolt one (paid by different distribution utilities) to entice farmers to make a switch to SIPs. A mechanism should also be put in place for distribution utilities to be compensated for the higher tariffs. In determining a higher bulk tariff, BERC could consider a rollback to bulk electricity tariffs after the first 5 years, by which time farmers and project sponsors will have recouped much of their initial investments. However, in the future, there should be provisions for these SIP systems to use grid electricity at off-peak hours while the priority setting of the solar pump energy inputs will be from solar photovoltaic. After 2030, all irrigation pumps should be connected to the main grid and be able to operate in hybrid mode, with solar as priority. Beyond the sunny periods, if there is need for electricity that is in surplus within the grid, the pumps should be enabled to use that. The *Guidelines for the Grid Integration of Solar Irrigation Pumps 2020* should be revised to allow import of electricity from 12 midnight to 7:00 a.m. or any other off-peak times of the day or year. This will also reduce dependency on standby diesel pumps by individual farmers.

For the case of dual solar-electric systems (dual systems can operate with solar energy during sunshine hours and with grid electricity on cloudy days or during night hours), compensation may occur through the net metering policy adopted by the Government of Bangladesh. This road map recommends that farmers with current electric pumps in need of replacement be offered the possibility of switching to dual solar-electric SIP systems and be allowed to operate under a net metering scheme. Under the process of net metering, electricity can flow in both directions via a bidirectional meter. The net energy, which is the total energy taken from the network minus the total energy supplied to the network during the designated billing period, is used to compute the customer's bill. This information is recorded on the meter. The consumer is responsible for paying the bill for net energy used if the amount of electricity consumed from the grid exceeds the amount provided to the grid. Conversely, in the event that the quantity of power produced and exported to the grid surpasses the amount provided by the grid, the distribution utility will allow the full amount of the customer's credit, measured in kilowatt-hours, to be carried over to the subsequent billing cycle. Any balance of electricity favorable to the consumer at the end of settlement period is paid to the consumer at a specified tariff. However, the current *Guidelines for the Grid Integration of Solar Irrigation Pumps 2020* does not foresee the application of the net metering scheme for solar pumps. This guideline will have to be revised to allow net metering in dual solar-electric systems. This road map also recommends that bulk electricity and net metering purchase tariffs offered to SIP system owners be frequently revised to ensure a reasonable return on investments by farmers to entice them to make the switch from diesel pumps to SIP systems.

Road Map Rollout

The road map is foreseen as being implemented in two phases. A dissemination phase during the period 2023–2026 will introduce solar irrigation in all suitable *upazilas* by installing the first 18,000 SIP systems; then, a market uptake phase from 2027 to 2031 will stabilize the market for SIP systems by installing an additional 27,000 SIP systems, aiming to reduce the level of subsidies in each loan wherever possible. The cost of SIP systems for Bangladesh would be $3,300 to $4,700 per kW pump installed; the total cost of implementing the road map is estimated at $1.8 billion. This will require public and private financing: a dedicated public fund (SIP Fund), as well as loans and grants worth $800 million from multilateral banks and bilateral partners as development assistance, and from the Green Climate Fund. Following the successful rollout of this phase of 45,000 SIP systems, the government could then consider accelerating SIP rollouts to 200,000 or 300,000 systems in the next phase from 2032 to 2040, building upon lessons learned.

Recommendations to Enable the Rollout

- **Institutional:** The Ministry of Power, Energy and Mineral Resources (MOPEMR) and BERC are in charge of assessing and proposing regulatory changes, such as a net metering scheme suitable for SIP systems and deciding on the convenience and timing for increasing bulk tariffs for the purchase of surplus electricity exported to the grid by SIP system owners. To coordinate the efforts of all the various implementing agencies, including BADC, BMDA, BWDB, DAE, distribution utilities, IDCOL, LGED, the Rural Development Academy (RDA), and others carrying out or financing solar irrigation projects with public funding, SREDA should be assigned a project coordination role. A separate national Technical Standard Committee should be established by SREDA for safeguarding the quality and standards of SIP systems that are being installed. Regarding the grid integration of SIP systems, all programs carried out under this road map shall leave the distribution utilities entirely responsible for evaluating and determining the suitability of grid connection on an individual basis.

- **Financing:** Conducive government policy supporting the cost reduction of equipment will be needed. This means considering preferential taxation, tax reductions or exemptions, or accelerated depreciation measures to mitigate the effects of high capital costs of renewable energy equipment in general, including SIP systems. This should be accompanied by policy changes that will enable grid connections of SIP systems with an appropriate net metering scheme.

- **Ownership and Capacity Building:** Targeted awareness campaigns about the benefits of SIPs and different financing models are recommended to increase ownership by farmers, enhance the role of women in farming activities, as well as increasing capacity building within different government agencies. These awareness campaigns can be undertaken by large businesses in Bangladesh that are committed to environment, social, and corporate governance in promoting SIPs as part of their media campaigns. Capacity building, in terms of enhancing skills in SIP O&M, is essential for the sustainability and scaling up of SIP systems. The primary objective of capacity building is to create a skilled workforce that will oversee the O&M of SIP installations. SREDA should take the lead in developing an adequate capacity building program that is aimed at

    - developing standardized maintenance practices at national level;
    - creating and distributing a variety of distance learning and classroom educational materials for all types of audiences such as technicians, master trainers, project developers, engineers, and policymakers;
    - establishing uniform training programs through a network of accredited training facilities throughout Bangladesh; and
    - creating a network of centers for entrepreneurship, technical training, and research and innovation to share best practices and encourage the transfer of knowledge.

- **Lowering the Cost Burden:** Surplus of electricity produced by SIP systems will be sold to the local distribution utility under the conditions established by the *Guidelines for the Grid Integration of Solar Irrigation Pumps 2020*. It is advised that these guidelines be updated as soon as possible to determine whether the import cap of 1 kilowatt-hour per kW alternating current (AC) is not too restrictive for the implementation of this road map. It is also recommended that a financing mechanism to subsidize the grid connection costs of existing (already installed) and new solar pumps to smaller and poorer farmers be established by the government, since this group of farmers will not be able to afford the extra financial burden of paying for the grid connection of their SIP systems. For community-based systems, provisions should be made to ensure that they are able to import electricity during cloudy days so as not to disrupt their irrigation needs. The same applies to individual farmers, but to a lesser extent—if they are not able to irrigate on a particular day, they could irrigate the next day without stressing the crop, since the soil water rootzone will provide the necessary buffer. Lastly, it is advised to put policies in place that will help farmers repay loans when their crops aren't doing well or when disasters like floods and cyclones strike. An illustration of such a measure would be agricultural product insurance.

- **Energy Storage for Grid-Connected Pump Systems:** Battery energy storage systems can be offered in a later road map (i.e., after 2031) as optional equipment for grid-connected SIP systems and for dual solar-electric pump systems. Energy storage could increase the availability of ancillary services in the grid. Specific regulation for energy storage and the provision of ancillary services will be needed for the full benefits of energy storage to materialize.

1. Introduction

Home to 171.68 million people and covering an area of 147,630 square kilometers, Bangladesh is a densely populated country with a fast-growing economy that aims to maintain a gross domestic product (GDP) growth above 6% beyond 2022.[1] To sustain this level of economic growth, the nexus between energy, water, and food at national level must be carefully planned and coordinated to keep transforming the country's economy in a sustainable way that benefits all citizens.

Bangladesh is a country primarily focused on agriculture, particularly rice farming. In addition to employing 40.6% of all workers and contributing 13.65% of the country's GDP, the agriculture industry is essential to reducing poverty and guaranteeing food security.[2] The agriculture sector's absolute GDP contribution is rising even though its percentage contribution is gradually decreasing. Bangladesh can now produce enough major cereals on its own. Today's vegetable production very nearly meets demand; fruits are being produced and exported in greater quantity and with higher quality.[3]

Bangladesh has three cropping seasons: Kharif I (mid-March–June), Kharif II (July–mid-October), and Rabi (mid-October until mid-March). Bangladesh receives abundant rainfall during the monsoon (June–October), but little rain from January to May, when Boro rice is cultivated.[4] Boro cultivation is supported by irrigation using forced mode pumps (submersible) and suction mode pumps (centrifugal).

Given Bangladesh's population, and to some extent uncontrolled groundwater irrigation practices, significant pressure is placed on the country's land and water resources, leading to pollution and depletion of shallow groundwater resources. This has detrimental impacts on food production and its costs. Evidence currently available indicates that resource development, as opposed to resource management, has received most of the attention in irrigation policy up to this point. Serious issues have arisen from this, the most significant of which being a drawdown (a decrease in the static water level) in heavily irrigated areas and a decline in the quality of groundwater.[5]

[1] Asian Development Bank. 2022. *Basic 2022 Statistics*. Manila. https://www.adb.org/sites/default/files/publication/788161/basic-statistics-2022.pdf.

[2] Government of Bangladesh, Bangladesh Bureau of Statistics. 2020. *Statistical Yearbook Bangladesh 2019. 39th Edition*. Dhaka. http://bcpabd.com/wp-content/uploads/2021/04/Statistical-Yearbook-of-Bangladesh-2019.pdf.

[3] Government of Bangladesh. 2020. *Bangladesh Voluntary National Reviews 2020: Accelerated Action and Transformative Pathways: Realizing the Decade of Action and Delivery for Sustainable Development*. Dhaka. http://file-rajshahi.portal.gov.bd/uploads/47d5d143-f8e7-4fb3-864b-45d66aad6c3f//636/c87/188/636c871882b91681133196.pdf.

[4] Boro rice is planted in winter and harvested in summer. Aman rice is sown during the rainy season (July–August) and harvested in winter.

[5] BADC. 2020. *Minor Irrigation Survey Report 2018–2019*. Dhaka. http://badc.portal.gov.bd/sites/default/files/files/badc.portal.gov.bd/page/c23bdffd_22fd_4f15_8fc4_b1fc7a91a36a/2020-09-01-14-15-bbb411c861df9a62fdafcccbd8025192.pdf.

In addition, Bangladesh is also highly vulnerable to the effects of climate change and will be impacted to a greater degree than most countries by 2025.[6] The effects of climate change on surface and groundwater resources are expected to be severe. Climate change vulnerability coupled with the country's high dependence on agriculture places significant pressures on the future availability of land and groundwater resources. Changes to water resources and hydrology will have a significant impact on agricultural irrigation, fisheries, industrial production, navigation, and other activities.[7]

Minor irrigation in Bangladesh is done with 1.22 million diesel pumps and more than 430,000 electric pumps. Irrigation equipment in Bangladesh is commonly classified as being shallow tube wells (STWs), deep tube wells (DTWs), or low-lift pumps (LLPs). STWs and DTWs pump groundwater, while LLPs pump surface water. The costs of irrigation in Bangladesh represent 43% of total agriculture costs.[8] Water efficiency must be increased and cropping patterns need to be rationalized, considering water availability and the sustainability of aquifers. Attention must be given to lessen pressure on groundwater. For example, Boro rice is fully irrigated, and Aman rice is partly irrigated; an estimated 3,000–5,000 liters (L) of water is required to produce 1 kilogram of rice. Alternate wetting and drying practices could save 15%–30% of irrigation water without reduction in rice yields. These good irrigation practices require organization and special care and contribute to improving the already fragile economy of farmers.

Solar energy is widely adopted in residential, commercial, and industrial sectors in Bangladesh and has huge potential to benefit the agriculture sector as well. Bangladesh has an average of 4–6 kilowatt-hours (kWh) per m^2 of solar radiation falling on its land over 300 days per annum.[9] The maximum amount of radiation is available in the months of March and April, and the minimum in December and January. The adoption of solar irrigation pump (SIP) systems to meet the irrigation requirements of farmers instead of diesel pumps is an opportunity to implement an appropriate and sustainable solution for Bangladesh. Modern SIP systems, especially those servicing a group of farmers, can be a central element in the improvement of groundwater management efficiency.

This road map proposes the solarization of irrigation services and the coexistence of solar panels and agriculture crops in the same area of land, a concept called agrophotovoltaics. This concept is being actively pursued in many developing and developed countries where land is constrained as part of their energy transition plans. This road map provides guidelines on how Bangladesh can manage a transition from diesel pumps to SIPs by 2031 by launching schemes that promote the installation of SIP systems and enabling their connection to the grid.

The general objectives of this road map are to

- harness environmentally friendly renewable energy sources and enhance their contribution to socioeconomic development in the rural context and also to increase the country's share of renewables;
- mitigate climate change effects by making more efficient use of groundwater and land resources and reducing carbon dioxide and greenhouse gas (GHG) emissions;
- put forward enabling policies and regulatory frameworks conducive to market development of SIP systems in farms that will result in further cost reductions and encouraging private sector participation;

---

[6] H. Ismail. 2016. *Climate Change, Food and Water Security in Bangladesh. Strategic Analysis Paper. Nedlands, Australia:* Future Directions International.

[7] Ministry of Foreign Affairs of the Netherlands. 2018. *Climate Change Profile of Bangladesh. The Hague.* https://www.government.nl/binaries/government/documenten/publications/2019/02/05/climate-change-profiles/Bangladesh.pdf.

[8] Government of Bangladesh, Ministry of Power, Energy and Mineral Resources. 2020. *Guidelines for the Grid Integration of Solar Irrigation Pumps 2020. Dhaka.*

[9] M.A. Hossain et al. 2015. *Solar Pumping System for Green Agriculture.* Agricultural Engineering International: CIGR Journal. 16. pp. 1–15. https://cigrjournal.org/index.php/Ejounral/article/view/2836.

- develop financing-working models for the long-term operation of SIP systems in farms, with an effective return on investment;
- develop sustainable delivery models that will result in scaling-up programs;
- create strong public awareness and involve users/local communities along with capacity building in establishing, operating, and managing SIPs in farms;
- support the government's commitment and ownership of SIPs through policy mechanisms that address the concerns of a wide range of stakeholders in meeting the high level of ambition of targeted SIPs; and
- create conditions for improving gender equality and providing opportunities for direct and indirect employment in rural areas.

By adopting this road map, the Government of Bangladesh will establish a national policy for the replacement of diesel pumps with SIP systems. Financing priority will be given to projects targeting irrigation with surface water.

1.1   Concerns Identified

The agriculture sector needs efficient and sustainable irrigation facilities and other amenities to reap proper benefits. The proposed road map has been built in coordination and consultation with several agencies of the Government of Bangladesh. These include the Sustainable Renewable Energy Development Authority (SREDA), the Power Division of the Ministry of Power, Energy and Mineral Resources (MOPEMR), Bangladesh Agriculture Development Corporation (BADC), Barind Multipurpose Development Authority (BMDA), Bangladesh Rural Electrification Board (BREB), Bangladesh Water Development Board (BWDB), Infrastructure Development Company Limited (IDCOL), Department of Agriculture Extension (DAE), and Rural Development Academy (RDA). The following concerns have been raised during consultations and have been duly addressed in the road map:

(i)   **Impact on agronomy and hydrogeology:** The road map should identify areas where there are opportunities for introducing SIP systems matching with current agricultural activities and needs in Bangladesh.

(ii)   **Institutional concerns:** The road map should identify policy deficiencies (gaps) and plausible financing-working models, suitable for each *upazila* (subdistrict), to scale up SIP in Bangladesh.

    (a)   For vulnerable communities, the road map suggests assessing the applicability of models established by the BADC and BMDA with applicable grants of 65% or more. These solutions serve groups of more than 30 farmers organized in self-administered committees or cooperatives. These models should focus on the use of surface water.

    (b)   For wealthier communities, the road map recommends the application of the model established by IDCOL with sponsors charging a fee for service. This model is currently working with a maximum grant of 50%.

    (c)   The road map suggests assessing the applicability of the model established by BREB where grants of 60%–65% are offered to individual applicants, and the rest is paid by each applicant over a 10-year period or extending the repayment period to 12–14 years when it may be necessary to lower monthly repayments.

The road map recommends that the MOPEMR's Power Division and the Ministry of Agriculture identify the capacities, strengths, and weaknesses of all candidate implementing agencies and institute capacity building of these agencies to deliver the SIP systems.

1.2 Overall Approach

The Asian Development Bank (ADB) has deployed a team of water and energy specialists to create this road map. The team has conducted an extensive literature review, consulted key stakeholders (government, private sector, and farmers), and performed multi-criteria analysis to address all identified concerns.

The first decision made was on the spatial scale of the analysis. The two levels the team considered in this road map are district and *upazila* (subdistrict). An abundant supply of open-source data is available at the district level, but this is quite coarse for the purpose of this exercise. The next level is at the *upazila* level, for which there is an abundant supply of data in publications. Based on the literature, the road map specialists have assessed

(a) whether there will be any increase in crop water requirements due to the rollout of the SIP road map,
(b) areas where groundwater levels are declining,
(c) available irrigation sources and priority locations for surface water and groundwater irrigation,
(d) the state of solar irrigation in Bangladesh,
(e) the institutional framework, and
(f) suitable irrigation schemes for Bangladesh.

Finally, the team proposes a rollout plan for the road map and a series of recommendations to enable this.

2. Road Map Scope

2.1 Objective

This road map aims to install approximately 45,000 SIP systems to replace approximately 200,000 diesel pumps, irrigating about 400,000 hectares (ha) of land and serving more than 1.3 million farmers.[10] These SIP systems would displace the consumption of 300,000 tons of diesel fuel annually, avoiding 900,000 metric tons of carbon dioxide equivalent (MtCO$_2$e) every year. They would add up to 1,000 megawatt peak (MWp) of solar capacities to the country, and also result in savings of $377 million annually, based on current diesel prices.[11]

These targets are significantly more ambitious than the interventions foreseen in agriculture by Bangladesh's Nationally Determined Contribution (NDC) and may require a consortium of donors, including ADB, to raise an estimated amount of $800 million in grants and loans to help implement the road map.

SIP systems offer the possibility of exporting renewable electricity to the national grid, since pumps operate on a full-time basis only for 100–150 days per year.[12] The new combined solar capacities will have an export capacity of electricity surplus to the grid of up to 480 gigawatt-hours (GWh) per year, equivalent to almost 1% of Bangladesh's total current electricity generation. The export of electricity to the grid is possible through the provisions of Bangladesh's *Guidelines for the Grid Integration of SIPs 2020*.

The surplus of electricity produced by SIP systems during off season could also be used in other agricultural activities such as threshing, winnowing, hulling, and milling.

The implementation of the road map will result in the empowerment of farmers by enabling them to sell their surplus renewable electricity to the grid while also reducing their dependence on diesel for irrigation activity and promoting a more sustainable and efficient use of groundwater resources.

2.2 Benefits of Solar Irrigation Pump Systems

Farmers traditionally adopted diesel pumps when grid access was unavailable. Diesel pumps are easy to move from one plot of land to another that needs irrigation. Despite their low initial cost, they only have a short life span—between 8,000 and 35,000 hours—depending on operating circumstances, maintenance practices, and the quality of the diesel engine and installation. According to standard operating procedures, they need a minor

---

[10]   Efficient management of groundwater resources (in STWs), coupled with a sensible grouping of farmers and an adequate selection and design of replacing SIP systems, will make it possible that each SIP system replaces an average of three to eight diesel pumps. The installation of 45,000 SIP systems could phase out the operation of up to 20% of existing diesel pumps.
[11]   The diesel price on 27 February 2023 was Tk109 per liter.
[12]   ADB consultants' estimate.

service every 250 hours and a major service every 500–1,000 hours. Since most diesel pumps need to be started by hand, remote pumping installations are more expensive to run.

SIP systems, on the other hand,

- are supplied with reliable, clean energy and do not pollute water, soil, or air with diesel spills and GHG emissions;
- are a perfect match for irrigation since water is needed mostly when solar irradiation is strong;
- may contribute to a more efficient use of groundwater resources, since these systems can be designed for servicing individual farmers or groups of up to 60 farmers, leading to a professionalization of irrigation practices;
- are a mature technology, easy to install, and operate autonomously; and
- have a higher initial cost but low operation and maintenance (O&M) costs. SIP systems can achieve a payback of 2–5 years over diesel pumps, depending on pump size.

The benefits that SIP systems offer in terms of cost, technical service, O&M, and environmental-related problems are summarized in Figure 1.

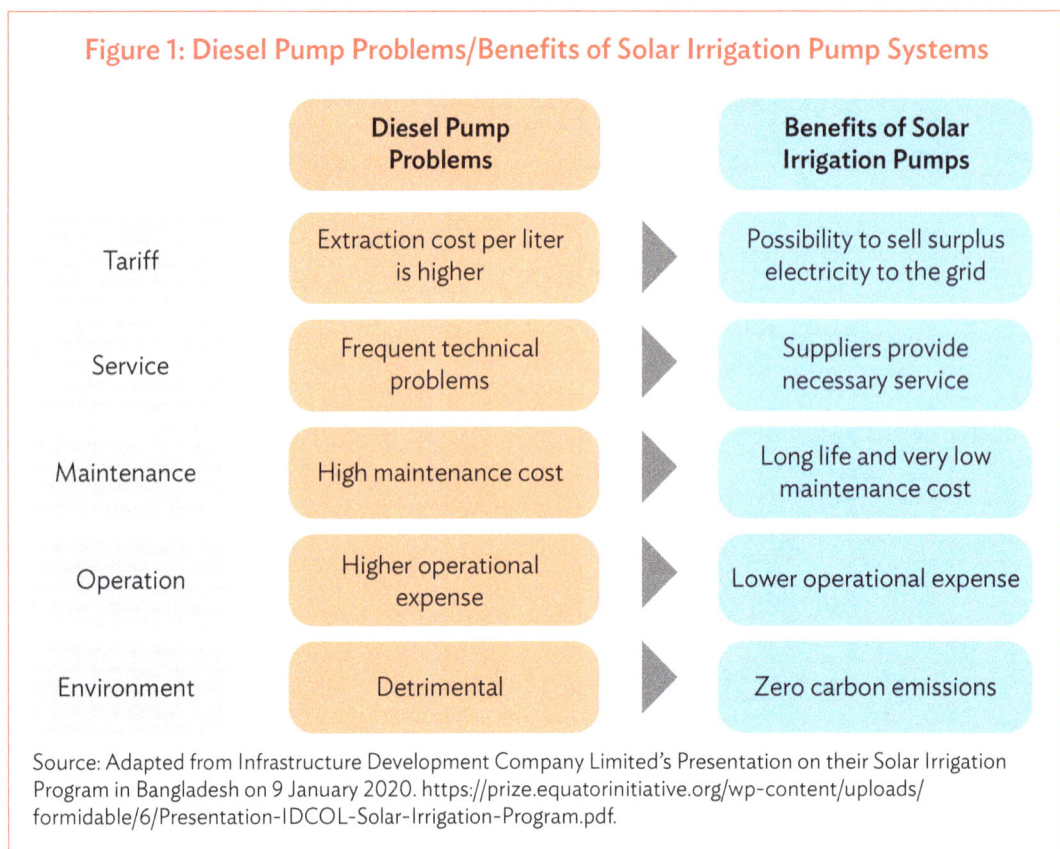

Figure 1: Diesel Pump Problems/Benefits of Solar Irrigation Pump Systems

	Diesel Pump Problems	Benefits of Solar Irrigation Pumps
Tariff	Extraction cost per liter is higher	Possibility to sell surplus electricity to the grid
Service	Frequent technical problems	Suppliers provide necessary service
Maintenance	High maintenance cost	Long life and very low maintenance cost
Operation	Higher operational expense	Lower operational expense
Environment	Detrimental	Zero carbon emissions

Source: Adapted from Infrastructure Development Company Limited's Presentation on their Solar Irrigation Program in Bangladesh on 9 January 2020. https://prize.equatorinitiative.org/wp-content/uploads/formidable/6/Presentation-IDCOL-Solar-Irrigation-Program.pdf.

The lifetime of SIPs is the same as that of conventional pumps: Submersible pumps last for 5–8 years and centrifugal pumps last for more than 30 years. Photovoltaic (PV) panels can last more than 25 years, while modern inverters can last at least 10 years. The intermittency of solar radiation is not a large problem, since energy produced can be stored in the form of water storage in overhead tanks and batteries, and then used in times when there is limited sunshine, or in the evenings for other purposes (washing and drinking water) in areas where there is no potable water. They are also more insulated from international diesel price fluctuations.

Many farmers have switched to electric pumps since the country achieved 100% electrification. Farmers in Bangladesh currently pay Tk4.19 per kWh compared to an average of Tk7.49 per kWh for residential, Tk8.53 per kWh for small industries, and Tk9.27 per kWh for commercial. These subsidized electricity tariffs have made the change to electric pumps more attractive to farmers. However, electric pumps frequently experience reliability issues due to poor quality of power supply. A low voltage on distribution lines often prevents farmers from operating electric pumps. Consequently, many still operate their diesel pumps when the electricity supply is not sufficiently reliable. Moreover, many electric pumps do not comply with the Groundwater Ordinance, 2019 and are unlicensed. The current low electricity tariffs will eventually be phased out and increase to reflect the real cost of supply. Electric pumps will then no longer be as attractive an option for farmers, but SIPs will.

An alternative to this situation is the installation of dual solar-electric systems that can operate with solar energy during sunny periods and with grid electricity on cloudy days or during night hours. Dual solar-electric systems provide reliable, predictable, and affordable energy for irrigation and increase the share of renewable electricity in the grid when the surplus is exported through the existing net metering scheme.

In relation to drinking water, SIP systems provide a valuable opportunity to increase access to safely managed, clean drinking water for the rural population. According to the Bangladesh Bureau of Statistics' 2019 Multiple Indicator Cluster Survey, just 47.9% of the country's population—51.2% of whom live in urban areas and 48.8% in rural areas—has access to safely managed drinking water. Bangladesh's target for 2030 is to have 100% of its population supplied with access to clean drinking water. The proportion of the population using safely managed drinking water services is measured by the proportion of the population using an improved[13] basic drinking water source, which is located on-premises,[14] available when needed,[15] and free of fecal (and chemicals including arsenic) contamination.[16] SIP systems can be combined with other technologies such as reverse osmosis and ultraviolet (UV) sterilization to produce drinking water and ozone sterilizers to remove bacteria and salinity in water. They can also be combined with ion-exchange resins or other filters to remove arsenic and other heavy metals to purify contaminated water. These technologies are very appropriate for the coastal areas of Bangladesh.

SIP systems can also help improve the livelihoods of many women who must travel long distances to obtain and transport diesel fuel for the pumps, or who are engaged in labor-intensive and time-consuming water fetching. In the long run, SIP systems and irrigation technology will contribute to closing the gender gap in the areas of economic resources, sufficient education, and access to ownership and control of land, inheritance, financial services, and natural resources, all in compliance with national legislation.

[13] Improved drinking water sources include piped water into a dwelling, yard, or plot; public taps or standpipes; boreholes or tube wells; protected dug wells; protected springs; packaged water; delivered water; and rainwater.
[14] A water source is located on-premises if the point of the collection is within the dwelling, yard, or plot.
[15] "Available when needed" means households can access sufficient quantities of water when needed.
[16] Government of Bangladesh. 2020. *Bangladesh Voluntary National Reviews 2020: Accelerated Action and Transformative Pathways: Realizing the Decade of Action and Delivery for Sustainable Development*. Dhaka.

2.3   Integrated Policy Approach

Water, energy, and food security are inextricably linked in Bangladesh. Energy acts as an enabler of increased food security, agricultural productivity, improvement of farmers' income, and improved access to water resources. Bangladesh's policy to reduce the use of shallow groundwater for irrigation, coupled with modernization efforts in agricultural production, and in particular the replacement of diesel pumps and hybridization of grid-connected pumps with efficient SIP systems, will have several positive environmental and economic impacts. The elaboration of this in the road map is a result of an integrated policy approach that weaves together different national policy objectives, such as the following:

- **Reduction of diesel fuel consumption and imports and GHG emissions.** The displacement of diesel fuel by solar energy reduces the national diesel fuel consumption and imports and makes irrigation a GHG emission-free activity. It also saves much-needed foreign exchange via importing less diesel (about $377 million annually) and the government saves on subsidized electricity provided to the agriculture sector.
- **Reduction of grid electricity consumption.** Bangladesh has 433,246 grid-connected irrigation pumps,[17] which require more than 2,000 megawatts (MW) of power from the grid between October and March, and consume more than 2,000 GWh per year of electricity. This challenges the grid during peak periods as additional expensive-generating capacities are needed to meet power demand.
- **Increased share of renewable energy.** By replacing large numbers of diesel pumps, several hundred MWp of intermittent clean renewable electricity is made available for export to the national grid at SIP idle times. To make this a viable solution to farmers, the government needs to establish a buyback electricity tariff that will encourage farmers to make the switch from diesel pumps to SIPs by showing them the quick payback period for their investments.
- **Eradication of food insecurity and poverty.** Bangladesh has achieved impressive gains in reducing food insecurity and poverty in recent years. But there is still more to do. The key priority of the country's National Food Policy is to increase the quantity, diversity, and quality of food produced. The use of modern and efficient irrigation techniques contributes to the extension of irrigation coverage in a sustainable and affordable manner and gives the possibility of earning additional income to farmers by selling their surplus electricity to the grid.
- **Better adaptation to climate change.** SIP infrastructure can be designed bearing in mind the possible threats of extreme weather, flooding, and cyclones when power lines are down. SIPs operate when there is sunshine and do not depend on the power grid. Adaptation measures to avoid irrigation disruption can therefore be better organized and implemented.
- **Sustainable use of groundwater resources.** There is a growing global consensus about the critical importance of groundwater for sustainable development and climate change adaptation. The World Economic Forum has stressed for the first time that depletion of this critical resource is causing megacities to sink, with significant risks to water security and resilience, and also threats to food production systems.[18] Modern and efficient SIP systems serving groups of farmers are an opportunity to improve the efficiency and management of groundwater to meet future irrigation demand.

---

[17]   BADC. 2022. Minor Irrigation Survey Report 2020-2021. Dhaka.
[18]   World Economic Forum. 2019. *The Global Risks Report 2019, 14th Edition. Geneva.* http://www3.weforum.org/docs/WEF_Global_Risks_Report_2019.pdf.

2.3.1   International Climate Commitments

The proposed road map is fully aligned with Bangladesh's NDC, as well as with its national energy and climate policies. Bangladesh submitted its first NDC report to the United Nations Framework Convention on Climate Change (UNFCCC) in September 2015. An updated NDC was submitted in August 2021. Bangladesh pledges in its NDC to contribute to the worldwide effort to lower future greenhouse gas emissions by boosting energy efficiency and the share of renewable energy in its national energy mix. Supported by the existing strategies and plans at that time (e.g., the Bangladesh Climate Change and Strategic Action Plan, Renewable Energy Policy 2008, Energy Efficiency and Conservation Master Plan, Vision 2021, and consecutive five-year plans), the NDC establishes unconditional and conditional GHG reduction targets for the industry, transport, and power sectors, along with additional mitigation measures in other sectors. Bangladesh's contribution is comprised of the following:

- An unconditional 27.56 million $MtCO_2e$ reduction in GHG emissions (6.73%) compared to business-as-usual levels by 2030, distributed as 26.3 million $MtCO_2e$ from the power sector, 0.64 million $MtCO_2e$ from agriculture, forestry, and other land uses, and 0.6 million $MtCO_2e$ from the waste sector. No reduction from industrial processes and product use is expected.
- An additional and conditional 34.34 million $MtCO_2e$ reduction in GHG emissions to achieve a total reduction of 61.9 million $MtCO_2e$ (15.12%) compared to business-as-usual levels by 2030 in the respective sectors.

The international commitment of Bangladesh is to achieve its unconditional goals with existing resources, while achieving its conditional goals is subject to appropriate international support in the form of loan and grant financing, investment, technology development and transfer, and capacity building.

According to the updated NDC, achieving the unconditional emission reduction goals would require the implementation of renewable energy projects of 911.8 MW, including 581 MW of grid-connected solar projects. Achieving the conditional emission reduction target would require the implementation of renewable energy projects of 4,114.3 MW, including 2,277 MW of grid-connected solar. However, the current estimated renewable energy capacity is 949 MW, so there is a need to accelerate investments in this subsector.

The NDC sets out several mitigation actions that will help reduce the country's GHG emissions. An "aggressive" scale-up of SIPs for irrigation is considered within the unconditional measures. Thus, the road map is fully aligned with the ambitions established in Bangladesh's NDC. The reduction of GHG emissions for the energy sector established by the NDC is achievable only when relevant parts of the potential for solar rooftop (residential, commercial, and industrial), floating solar PV systems, ground-mounted solar PV systems, SIP systems, and other renewable energy sources (wind and biomass) are realized. The operation of diesel pumps currently consumes 1 million tons of diesel per year (estimated at $1.26 billion at 2022 diesel prices) and emits 3 million $MtCO_2e$ per year. The transition of diesel-based irrigation to solar irrigation under this road map will not only reduce 900,000 $MtCO_2e$ emissions during irrigation but also offer the possibility to inject the excess electricity produced into the national grid, thus displacing power produced with fossil fuels. For the implementation of unconditional measures, the NDC and Bangladesh Climate Change and Strategic Action Plan estimate an allocation of $3.25 billion will be required.

2.3.2   Energy Policies

By 2020, the government had aimed to raise the percentage of renewable energy to 10%, but the goal was not achieved. SREDA reported that the installed renewable energy generation capacity in 2022 was 3.71%. By 2030, the Power Division wants to increase its capacity for power generation to 40,000 MW. The government's Mujib

Climate Prosperity Plan—Decade 2030 sets a 40% renewable energy target by 2041 (equivalent to approximately 40,000 MW of renewable energy generation capacity) and states that domestic energy security will be driven by investment in solar PV. Achieving this target will only be possible if investments in solar rooftop (residential, commercial, and industrial), floating solar PV systems, ground-mounted solar PV systems, and SIP systems are realized. A shift to renewables would help the country to reduce power subsidies or tariff increases as well. Along with commitment to solar, a focus on power grid upgrades will also be required. This will give the power system increased capacities of renewable energy and reduce transmission and distribution losses.

Scaling up the conversion of many diesel pumps to solar PV pumps would result in a few hundred MW of intermittent and seasonal readily available PV electricity capacity that can be connected to the national grid. The Mujib Climate Prosperity Plan—Decade 2030 recommends studying a displacement strategy for technologically outdated fossil fuel based power plants toward new technologies, and explicitly mentioning the co-location of SIP systems in farming areas to replace diesel powered irrigation pumps. The plan not only recognizes the national need for a major upscaling of SIP systems, but also that power from these systems can be fed to the national grid, which means that a significant amount of renewable energy capacity addition will be realized with the implementation of this road map.

The integration of solar power into the grid would mean a significant reduction in the country's national emissions. The support mechanism for selling electricity to the grid should be built on the provisions given by the *Guidelines for the Grid Integration of Solar Irrigation Pumps 2020*. This mechanism should enable the sale of surplus electricity generated by SIP systems to the grid. The current price set by the government is not attractive enough for farmers to switch from diesel pumps to SIPs. The road map recommends that the government considers a higher price to entice farmers to make the switch. In determining the buyback rate, the government should consider the contribution SIP systems will make in achieving the country's national renewable energy targets and make the payback of the project to farmers more attractive. In this regard, Bangladesh Energy Regulatory Commission (BERC) could consider offering a higher tariff than the current 33 kilovolt (kV) bulk electricity tariff for the first 5 years, and later dropping to the bulk tariff rate for the remainder of any power purchasing agreements with distribution utilities.

Appropriate support policies are expected to be enacted after the Integrated Power and Energy Master Plan is published by the Government of Bangladesh in 2023. Those future support policies should aim to address market failures in promoting the uptake of renewable energy while achieving a number of other objectives, including energy diversification, energy security, the development of a local industry, and job creation.

2.3.3   Agricultural and Water Policies

The proposed road map is fully aligned with the following national agricultural policies:

- **Eighth Five-Year Plan:** The Eighth Five-Year Plan on agriculture recognizes that solar irrigation contributes to farm mechanization and modernization. The SIP road map is also fully aligned with these policy objectives. The replacement of uncontrolled and not sufficiently regulated small diesel pumps with modern and efficient SIP systems serving groups of farmers, rather than individual fields, is a pillar of Bangladesh's sustainable irrigation strategy to manage groundwater resources more efficiently, reduce irrigation costs, and improve farmers' income.
- **National Agricultural Policy 2018 and Agricultural Expansion Strategy:** Bangladesh's agricultural policy is to increase efforts to modernize agricultural production through mechanization, innovation, technology transfer, and agricultural extension. It also aims at reducing the depletion of groundwater by reducing and managing its use more effectively. The road map is aligned with these policy objectives.

The use of SIP systems serving groups of farmers is an innovation that improves irrigation access and food security to a larger share of Bangladesh's rural population. The implementation of the road map also reduces irrigation costs, manages groundwater resources more effectively, and reduces the contamination of groundwater and soil.

- **Minor Irrigation Policy 2017:** The road map is aligned with the Minor Irrigation Policy goals of developing agriculture to increase production, food security, and promote efficient irrigation and efficient groundwater management. The proposed financing-working models are also aligned with the policy goal of promoting associative and cooperative approaches in irrigation activities. The road map recommends that all government-supported programs should first evaluate the possibility of changing the irrigation configurations from STWs to surface water using LLPs to reduce the use of groundwater, as established by the Minor Irrigation Policy. Where surface water is not possible, the option of DTW configurations should be explored as a better alternative to STWs.
- **Groundwater Management Rules 2019:** The provision and guidelines given by this road map are consistent with the current management rules for the formation of the *upazilas'* Irrigation Committees, the application of licenses for the installation of irrigation pumps, the requirements of irrigated area and maximum water flow, as well as the provisions for the supervision of approved licenses.
- **National Agricultural Mechanization Policy 2020:** This policy includes a target of 10% use of renewable energy in agricultural activities by 2020 and 50% by 2050.
- **Water management policies:** The road map is aligned with water management policies as it recommends that all STW projects should include the obligation to test for arsenic contamination and groundwater depletion as requisites to approve an irrigation/drinking water project. The road map establishes that such tests should be monitored by the DAE and Local Government Engineering Department (LGED).

2.4   Contribution to Sustainable Development Goals

The most significant impact is that replacing diesel pumps with more advanced and effective SIP systems will affect two Sustainable Development Goals (SDGs): combating climate change (SDG 13), and providing affordable and clean energy (SDG 7). Improved use of renewable energy is a result of the introduction of SIP technology through suitable financing mechanisms in Bangladesh. When compared to diesel water pumps, SIPs provide a longer-term, more economical, and cleaner system.

2.4.1   Clean and Affordable Energy

The vision of the Power Division of the MOPEMR is to achieve the SDGs and boost Bangladesh's economy to the level of an upper-middle-income country by 2031 through making affordable and dependable electricity available to all.[19] Bangladesh has already achieved a 100% electrification rate.

The replacement of diesel pumps with modern SIPs is not only the best option for achieving reliable, clean energy and affordable irrigation services, but it also provides the national power grid with a large amount of ready-to-be-connected installed capacity from solar panels that can export its excess electricity production during idle agriculture times to the grid.

---

[19]   Government of Bangladesh. 2020. *Bangladesh Voluntary National Reviews (VNRs) 2020: Accelerated Action and Transformative Pathways: Realizing the Decade of Action and Delivery for Sustainable Development.* Dhaka.

The power sector in Bangladesh is characterized by recurring power outages of 8–12 hours in rural areas due to extending long distribution lines to increase the country's electrification rate. This causes a voltage drop along distribution lines resulting in power outages and faults in the lines, with rural households not being able to operate their appliances and electric pumps. The distribution and high-voltage transmission network needs to be strengthened. Power outages also often occur in urban areas for 10–12 hours daily, due to imports of liquefied natural gas (LNG) being reduced on account of high international prices. Power outages constitute a critical constraint to Bangladesh's internal and external competitiveness and jeopardize inclusive growth. The total on-grid installed generating capacity as of January 2023 was 25,782 MW, of which only 606.53 MW is on-grid renewable energy. This is only 2.4% of the total on-grid installed capacity. Hydropower accounts for 230 MW, solar for 375.63 MW, and wind for 0.9 MW. Additionally, 2 MW of wind energy and 356.86 MW of solar energy have been installed in off-grid solar home systems.[20]

The country needs to increase its share of renewable energy in generation from 950 MW in 2022 to 4,500 MW by 2030 to reduce GHG by 6.73%, in line with the country's commitment in its 2021 updated NDC.

2.4.2 Action to Combat Climate Change

Bangladesh, which already struggles with poverty and environmental degradation, is under stress from climate change, which has a particularly negative effect on people living in rural areas. The demand for irrigation will rise as a result of the possible effects of climate change, necessitating the construction of more durable irrigation infrastructure.

Bangladesh's landscape is mainly flat. A large part of the country comprises alluvial plain, created by the two great river systems of the Ganges (Padma) and Brahmaputra (Jamuna), and their innumerable tributaries.[21] Its geographical location makes it particularly susceptible to extreme weather events, including cyclones, floods, and storm surges. According to the 2017 Notre Dame Global Adaptation Initiative index, which summarizes a country's vulnerability to climate change and other global challenges in combination with readiness to improve resilience, Bangladesh is the 33rd most vulnerable country and the 25th least-ready country,[22] meaning that while it is highly vulnerable, it is not ready to prevent or reduce climate change effects.[23] Bangladesh has a long history of disasters, experiencing 219 between 1980 and 2008 that caused over $16 billion in total damage. On average, a powerful cyclone strikes the nation every 3 years.[24]

According to studies, the effects of climate change reduce agricultural GDP output by 3.1% annually, which amounts to a $36 billion loss in added value between 2005 and 2050. If indirect effects on complementary industries are taken into account, this rises to a total of $129 billion.[25] In a low productivity scenario, climate change could result in a net increase in poverty of up to 15% between 2000 and 2030.[26] In addition, poverty and food insecurity make people less resilient and less able to adapt to the effects of climate change, making women and girls especially susceptible to its consequences.

[20] SREDA. National Database of Renewable Energy. https://ndre.sreda.gov.bd/index.php?id=4 (accessed November 2022).

[21] J. Ayers et al. 2014. Mainstreaming Climate Change Adaptation into Development in Bangladesh. *Climate and Development*. 6(4). pp. 293–305. http://dx.doi.org/10.1080/17565529.2014.977761.

[22] The country's susceptibility to the adverse impacts of climate change in six essential areas for human survival—food, water, ecosystem services, health, human habitat, and infrastructure—is measured by vulnerability. The concept of readiness evaluates a nation's capacity to transform investments into adaptation measures by taking into account its level of social, political, and economic preparedness.

[23] ADB. 2020. *Asian Water Development Outlook 2020*. Manila.

[24] Government of Bangladesh, Ministry of Environment and Forests. 2009. Bangladesh Climate Change Strategy and Action Plan. Dhaka.

[25] World Bank. 2010. *Economics of Adaptation to Climate Change: Bangladesh*. Washington, DC. https://openknowledge.worldbank.org/bitstream/handle/10986/12837/702660v10ESW0P0IC000EACC0Bangladesh.pdf?sequence=1.

[26] H. Wright. 2014. What Does the IPCC Say about Bangladesh? *International Centre for Climate Change and Development Briefing*. https://www.eldis.org/document/A70952.

Every region will experience climate change in a different way, as follows:

- The hardest hit areas from drought and rising temperatures will be the northwest.
- Floods will occur more frequently and with greater intensity in the center and northeast.
- Sea level rise, saline intrusion, and increased cyclone frequency and intensity will affect all coastal areas and islands, with urban coastal areas particularly suffering from drainage congestion.[27]

In 2010, farmers reported that, on average, shocks cost them 12% of their harvests; half of these shocks were caused by flooding, which included river erosion and waterlogging (footnote 27). Floods originate from precipitation in the whole of the Ganges–Brahmaputra–Meghna river basin, not just the 7% that lies within Bangladesh, and can therefore be of great magnitude (footnote 25). Almost every year floods occur in July and August, with on average, about 25% of the country inundated.[28] Severe floods with devastating effects cover 60% of the country every 4–5 years. Thousands of hectares of agricultural land have been lost as a result of riverbank erosion caused by floods (footnote 24) and has long-term effects on the affected population (footnote 29).

In addition to causing harvest losses, floods have also added to the salinization of coastal areas, resulting in the loss of productive agricultural land (footnote 27). About 1.2 million ha of arable land out of 2.85 million ha of coastal and offshore areas are already impacted by varying degrees of soil salinity problems.[29] As climate change continues, these effects are probably going to get more severe (footnote 26). According to one 2017 study, the salination of land in Bangladesh could cause a 15.6% decline in rice yields.[30] While flooding is common and widespread in many parts of Bangladesh, there are also occasional seasonal droughts in other areas.[31] This occurs especially in the months leading up to the November–December rice harvest in the northwest of the country, where they may cause up to a 40% reduction in crop productivity by 2050 (footnote 24).

The effects of climate change on crops are relatively well documented. Some estimates state that Bangladesh's total rice production will have dropped by roughly 8% from 1990 to 2050; other estimates state that the effects of climate change will result in a cumulative loss of 80 million tons of rice between 2005 and 2050 (footnote 6). The potential impacts of climate change on the most important crops are as follows:

- **Aman and Boro rice:** More frequent and severe flooding will have an impact on Aman rice production, while Boro rice will be negatively impacted by a (seasonally) limited supply of surface water and decreasing groundwater levels. By 2030, seasonal drought will affect about 60% of the land used for rice production, which means that a comparable percentage of Bangladesh's rice yields will be impacted by drought in one way or another. The southern part of the country is most at risk: By 2050, production losses of 10% for Aman rice and 18% for Boro rice are anticipated in the Khulna region, primarily as a result of sea level rise (footnote 28).

27 T.S. Thomas et al. 2013. Agriculture and Adaptation in Bangladesh: Current and Projected Impacts of Climate Change. *IFPRI Discussion Paper. No. 01281*. Washington, DC: The International Food Policy Research Institute. http://www.ifpri.org/sites/default/files/publications/ifpridp01281.pdf.

28 Z. Sharmin and M.S. Islam. 2013. *Consequences of Climate Change and Gender Vulnerability: Bangladesh Perspective. Bangladesh Development Research Working Paper Series*. BDRC: Dhaka. http://www.bangladeshstudies.org/files/WPS_no16.pdf.

29 World Bank. 2011. *Vulnerability, Risk Reduction, and Adaptation to Climate Change: Bangladesh. World Bank Climate Risk and Adaptation Country Profile*. Washington, DC. https://climateknowledgeportal.worldbank.org/country/bangladesh/vulnerability.

30 Norwegian Institute of Bioeconomy Research. 2017. Food Security Threatened by Sea-Level Rise. *Phys.org*. 18 January. https://phys.org/news/2017-01-food-threatened-sea-level.html.

31 S. Xenarios et al. 2014. *Agricultural Interventions and Investment Options for Climate Change in Drought and Saline-Flood Prone Regions of Bangladesh*. Ås, Norway: Bioforsk. http://www.riceclima.com/wp-content/uploads/2014/05/BIOFORSK-RAPPORT_AgricInterventions.pdf.

- **Paddy rice:** By 2030, inundation will have destroyed 121,000 tons of paddy rice and about 55,000 ha of paddy land. Approximately 20% of the entire paddy cultivation area will eventually be covered by salinized soil, which could result in the loss of an additional 395,000 tons of rice.
- **Wheat:** Up to 32% of Bangladesh's total wheat production could be lost by 2050, mostly as a result of heat stress (footnote 6). About 15% of yields are predicted to be lost even with the best planting dates and techniques, such as creating heat-tolerant varieties and enhancing fertilizer applications. (footnote 27). Since up to 50% of their total wheat yields may be lost, it will be difficult and expensive to import wheat from other South Asian countries (footnote 26).
- **Potatoes:** As of 2023, 64% of yields are lost due to moisture stress produced by untimely rainfall; by 2030, this will rise to 76%. To deal with this, up to 22% more irrigation would be needed.[32]
- **Maize:** If the current planting calendar and varieties are used, a 10%–20% productivity reduction is anticipated between 2000 and 2050. Maize yields could rise if recently optimized dates and varieties are used. Given that maize prices are expected to rise by 209% by 2050, more than any other food commodity, some have suggested that Bangladeshi farmers would benefit from switching from producing rice to maize.
- **Sugarcane, soybeans, and sorghum:** Regardless of changes in farming practices, productivity of these rainfed crops is predicted to decrease by 7.5%–10% between 2000 and 2050 (footnote 27).

A significant advancement in tackling climate change, water stress, and land scarcity is solar irrigation. Initiatives centered around solar irrigation offer a novel, cutting-edge, and possibly workable way to combat some of the problems caused by climate change. Additionally, they present a chance to improve the design of more resilient infrastructure in order to lessen the damaging effects of catastrophic flooding.

2.4.3 Other Sustainable Development Goals

The implementation of this road map also has clear impacts on SDG 1 (no poverty), SDG 2 (zero hunger), SDG 6 (water and sanitation), SDG 8 (decent work and economic growth), and SDG 10 (reduced inequalities). Double the agricultural output and incomes of small-scale farmers, especially the bottom 40% of subsistence farmers, will help end hunger and malnutrition, combat poverty, lessen inequalities of farmers, and spur economic growth for farmers. Resilient agricultural approaches and sustainable food production systems will be essential to reaching this objective. Year-round agricultural growing is encouraged by SIPs. This translates into increased food production, more effective land use, and more profits for farmers. SIP systems, which replace diesel pumps, provide power to a segment of population that has historically been left out of financial products. It enables people to escape a vicious cycle of poverty, inequality, and food insecurity. Solar irrigation aims to stimulate agriculture, which is a vital industry for Bangladesh's economy. Investing in this area could lead to improved work conditions, social development, and economic growth in addition to ensuring food security. Access to energy and clean water is also facilitated by technologically advanced and effective SIP systems.

In relation to the use of groundwater, the key features relevant to the SDGs are its efficient management and sustainability. Although the SDGs do not mention groundwater explicitly, one report did find interlinkages between sustainable use of groundwater and targets of the SDGs.[33] The efficient and sustainable use and management of groundwater has clear links with SDG 6 (water and sanitation), SDG 12 (responsible consumption and production), and SDG 13 (climate action). Most of these links are

---

[32]   S. Xenarios et al. 2014. *Agricultural Interventions and Investment Options for Climate Change in Drought and Saline-Flood Prone Regions of Bangladesh.* Ås, Norway: Bioforsk. http://www.riceclima.com/wp-content/uploads/2014/05/BIOFORSK-RAPPORT_AgricInterventions.pdf.

[33]   L. Guppy et al. 2018. Groundwater and Sustainable Development Goals: Analysis of Interlinkages. *UNU-INWEH Report Series, Issue 04.* Hamilton, Canada: United Nations University Institute for Water, Environment and Health. https://inweh.unu.edu/wp-content/uploads/2018/12/Groundwater-and-Sustainable-Development-Goals-Analysis-of-Interlinkages.pdf.

"reinforcing," which means that achievement of one SDG target will have a predominantly positive impact on the others.

2.5   The Water–Energy–Food Nexus

Despite the temporal scarcity of water, Bangladesh is a water-abundant country. A population density of approximately 1,260 people per square kilometer makes Bangladesh the 10th most densely populated country in the world. High density of population means that agricultural land is virtually saturated, with limited capacity to expand food production. The population of Bangladesh is moving from a mainly starch-based diet to a diet incorporating more water-intensive meat and dairy. Bangladesh largely relies on imported diesel to secure its food supplies. Roughly 3.5 million $MtCO_2e$ in Bangladesh in a year is from diesel-based irrigation, which is 4.4% of total annual production-based carbon dioxide emissions.[34]

Climate change will have dramatic impacts on agriculture, resulting in flooding and drought due to weather changes and geopolitical influences on trans-border rivers. Rising sea levels and a consequent increase in salinity will affect crops and require shifts to alternative land use. Water-induced disasters devastate food supplies, while land acquisition for the setting up of power plants, along with their discharged effluents in rivers, cause damage to water quality, agriculture, and fisheries in particular locations. Despite Bangladesh's national plans and policies focusing on water, energy, and food security, a lack of considering and incorporating the water–energy–food nexus impedes ecosystem sustenance.

The introduction of a system of rice intensification and an alternative wetting and drying method of irrigation practices through SIP systems will contribute to achieving higher agriculture yields and reduce agricultural production costs in an environmentally sustainable manner. Agriculture can be viewed as a coupled social-environmental system in which farmers depend on market forces to determine how much money they can make, public policies that control their access to these resources (such as capital in the form of pumps), and environmental inputs (such as water, but also seeds, fertilizer, and sunshine).

To realize these benefits, water–energy–food nexus coordination and planning at the national and cross-sectoral level is required to introduce the SIP system technologies into the market. There are many feedback loops within this nexus: insufficient rains in one year could lead to more government assistance for farmers the following year; new irrigation pump subsidies could result in more land being farmed; and financial incentives for farmers to use less water for their crops could lessen overextraction of groundwater. The national policy should reinforce factors to curb groundwater usage, as well as improving agricultural extension services and providing access to low-cost financing to small farmers. Thus, policies that combine the replacement of diesel pumps with SIP systems with mechanisms to export excess electricity back into the grid help to minimize groundwater use. This will come to pass as farmers become increasingly motivated to irrigate their fields efficiently in order to maximize the amount of energy they can sell to the grid.[35]

[34]   A. Mitra, M. F. Alam, and Y. Yashodha. 2021. *Solar Irrigation in Bangladesh: A Situation Analysis Report.* Colombo: International Water Management Institute.
[35]   Comprehensive Initiative on Technology Evaluation. 2017. *Solar Water Pumps: Technical, Systems, and Business Model Approaches to Evaluation.* Cambridge, MA: Massachusetts Institute of Technology.

3. Irrigation Requirements

3.1 Suitable Crop Areas

3.1.1 Available Rainfall

Bangladesh experiences warm, humid summers and comparatively cold, dry winters due to its sub-humid to humid monsoon climate. The five monsoon months of June through October account for about 80% of the country's annual rainfall, which averages 2,666 millimeters (mm), with the northeast receiving about 5,680 mm and the west receiving 1,100 mm. The wet season from May to October delivers 90% of total annual rainfall, leaving the remainder to occur throughout the dry season from November to April.[36]

Estimating the effective rainfall across Bangladesh from the total rainfall during the Boro season (January to May) is necessary to determine the net crop water requirement in *upazilas*. As observed rainfall data is not readily available, monthly data (from 1981 to May 2022) collected by the Climate Hazards Group InfraRed Precipitation with Station for each *upazila* has been used. The long-term average rainfall across Bangladesh and for each *upazila* was estimated, along with seasonal changes and trends. The resulting distribution of effective rainfall across Bangladesh is presented in Figure 2.

3.1.2 Increase in Crop Intensity

While the cropped area in Bangladesh has been stable, the cropping intensity has been gradually increasing. This is because of the development and dissemination of new generation short-duration rice and non-rice cultivars that fit into the fallow period in the existing single two-crop-based cropping patterns. Some single-cropped areas have been transformed into double-cropped areas, some double-cropped areas into triple-cropped areas, and some triple-cropped areas to quadruple-cropped areas. The cropping intensity of 147% in 1969–1970 increased to 195% during 2016–2017, at a rate of 0.65% per annum. It is expected that this cropping intensity will reach 215% by 2030. The availability of other green revolution technologies, including chemical fertilizers, pesticides, and irrigation facilities, has also contributed to the increase in crop intensity.

[36] S. J. Ahammed, E. S. Chung, and S. Shahid. 2018. Parametric Assessment of Pre-Monsoon Agricultural Water Scarcity in Bangladesh. Sustainability. 10(3). pp. 819. https://www.mdpi.com/2071-1050/10/3/819.

Figure 2: Distribution of Effective Boro Rainfall, 1980–2021
(mm)

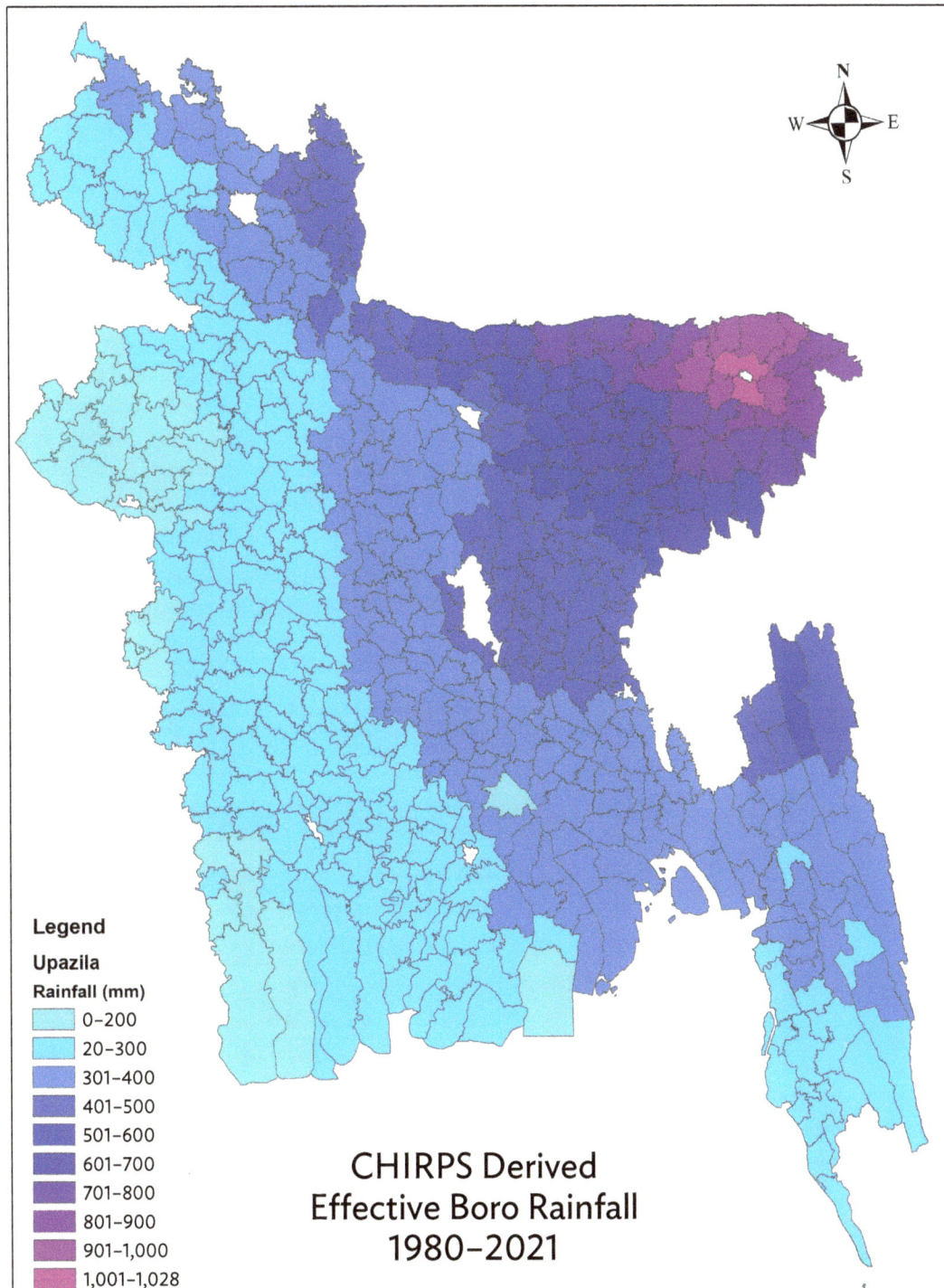

Legend

Upazila

Rainfall (mm)

- 0–200
- 20–300
- 301–400
- 401–500
- 501–600
- 601–700
- 701–800
- 801–900
- 901–1,000
- 1,001–1,028

CHIRPS Derived
Effective Boro Rainfall
1980–2021

CHIRPS = Climate Hazards Group InfraRed Precipitation with Station, mm = millimeters.

Source: Asian Development Bank consultants.

3.1.3   Evolution/Change of Irrigated Areas

While the total rice crop area has been reasonably steady in Bangladesh, the area under cultivation of other crops has been increasing. In 2020–2021, Bangladesh's net cultivable area was about 8.46 million ha. The Institute of Water Modelling mapped suitable areas for Boro irrigation. Their interim results show that 2.36 million ha are very suitable; 1.21 million ha are suitable; 0.53 million ha are moderately suitable, and 0.13 million ha are marginally suitable, totaling 4.23 million ha as suitable for irrigation during the Boro season. However, the area irrigated during Boro season 2020–2021 was 5.65 million ha. This means that all appropriate areas are likely under irrigation already, and further expansion is unlikely.

3.2   Distribution of Irrigation Pumps

In 2021, more than 5 million ha were irrigated, 72.65% of agricultural land was irrigated with pumped groundwater during Rabi season (October to March), and 27.35% of agricultural land was irrigated with surface water. In terms of energy source, 56.14% of agricultural land was irrigated with 1.22 million small itinerant diesel pumps, and 43.86% of agricultural land was irrigated with 0.44 million electric grid-connected pumps (Table 1). Diesel pumps in Bangladesh consume about 1 million tons of diesel per year ($1.26 billion in 2022 prices), emitting more than 3 million $MtCO_2e$.[37] The nation's 0.44 million electric irrigation pumps consume more than 2 GWh of grid power each summer.[38] Electricity is provided at a highly subsidized tariff to support farmers.[39] This creates an externality that distorts the true cost of agricultural production and happens at the expense of other agricultural innovations to increase productivity.

Table 1: Distribution of Irrigation Pumps in Bangladesh

Division	Electric		Diesel		Total	
	Number	Area (ha)	Number	Area (ha)	Number	Area (ha)
Dhaka	72,831	307,757	128,589	369,937	201,420	677,694
Mymensingh	58,647	312,405	112,617	276,488	171,264	588,893
Rajshahi	101,231	755,618	230,528	475,302	331,759	1,230,920
Rangpur	107,433	465,957	296,944	586,148	404,377	1,052,105
Chattogram	41,438	257,667	83,148	317,879	124,586	575,546
Khulna	42,122	190,063	279,904	549,566	322,026	739,629
Sylhet	11,037	60,854	61,033	287,374	72,070	348,228
Barishal	3,030	25,171	25,026	156,856	28,056	182,027
Total	**437,769**	**2,375,492**	**1,217,789**	**3,019,550**	**1,655,558**	**5,395,042**

ha= hectares.

Source: Bangladesh Agricultural Development Corporation. 2020. *Minor Irrigation Survey 2018–2019. Dhaka.*

---

[37] World Bank. 2015. *Solar-Powered Pumps Reduce Irrigation Costs in Bangladesh.* http://www.worldbank.org/en/results/2015/09/08/solar-powered-pumps-reduce-irrigation-costs-bangladesh.

[38] S. Islam. 2020. Bangladesh Net Metering Requirement May Deter Solar Pump Owners. *PV Magazine. August 6.* https://www.pv-magazine.com/2020/08/06/bangladesh-net-metering-requirement-may-deter-solar-pump-owners/.

[39] Bulk tariffs at 11 kV approved by the BERC in February 2020 range from Tk4.3679 to Tk6.4531 per kWh.

The distribution of pumps per type of irrigation equipment is shown in Table 2. STW pumps account for 86% of all diesel pumps in Bangladesh and irrigate 65% of all land irrigated with diesel pumps, while LLPs account for almost 14% of all diesel pumps and irrigate 34% of the land irrigated with diesel pumps.

Table 2: Major Electricity and Diesel Irrigation Pumps Operating in Bangladesh

Irrigation Equipment	Electricity Operated		Diesel Operated		Total	
	Units	Area Irrigated (ha)	Units	Area Irrigated (ha)	Units	Area Irrigated (ha)
DTW	35,059	1,052,147	1,896	33,284	36,955	1,085,431
STW	378,225	1,044,894	1,031,464	1,961,180	1,409,689	3,006,074
LLP	19,962	261,927	184,429	1,025,086	204,391	1,287,013
DTW+STW+LLP	**433,246**	**2,358,968**	**1,217,789**	**3,019,550**	**1,651,035**	**5,378,518**

DTW = deep tube well, ha = hectare, LLP = low-lift pump, STW = shallow tube well.

Data used and adapted from Bangladesh Agricultural Development Corporation. 2020. *Minor Irrigation Survey 2020–2021*. Dhaka.

Diesel pumps are mostly used in small, privately owned STWs irrigating about 2 million ha. Many farmers rent their diesel pumps and are subjected to higher fees during peak irrigation and crop harvesting seasons, which can cause them financial difficulties. The transportation of diesel to crop fields is also challenging and the supply of diesel can be inconsistent. A smaller number of farmers who are fortunate enough to own electricity-run pumps face persistent power outages. This forces them to operate their pumps at night when electricity demand is lower and power failures are fewer.

In the northwestern and central parts of the country, 89% of the irrigated area uses diesel pumps, which accounts for 78% of farmers. The northwestern part of the country hosts 67% of all diesel pumps used for irrigation in the country. Rangpur division (24%) has the largest number of diesel pumps, followed by Khulna (22%) and Rajshahi (22%). The central part of the country hosts 22% of diesel pumps—12% operating in Dhaka and 10% in Mymensingh. The eastern and southern parts of the country use fewer diesel pumps (Figure 3 and Table 3).

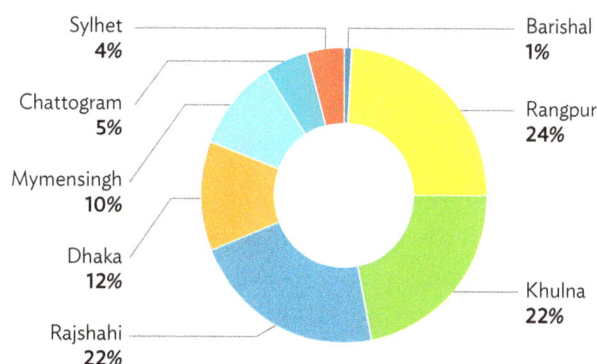

Figure 3: Distribution of Diesel Pumps across Bangladesh by Division

Sylhet 4%
Barishal 1%
Chattogram 5%
Rangpur **24%**
Mymensingh **10%**
Dhaka **12%**
Khulna **22%**
Rajshahi **22%**

Source: Data extracted and adapted from Bangladesh Agricultural Development Corporation. 2022. *Minor Irrigation Survey 2020–2021*. Dhaka.

Table 3: Distribution of Diesel Pumps per Division and Type of Irrigation Equipment

Division	Equipment	Units	Area Irrigated (ha)	Farmers
Rangpur	DTW	25	274	2,093
	STW	295,610	577,559	2,612,575
	LLP	1,309	8,315	16,189
Khulna	DTW	559	10,624	30,025
	STW	244,402	413,233	2,159,276
	LLP	34,943	125,709	280,527
Rajshahi	DTW	576	5,322	59,379
	STW	219,897	423,893	2,044,794
	LLP	10,055	46,087	98,990
Dhaka	DTW	195	4,121	14,504
	STW	106,633	209,063	815,903
	LLP	21,761	156,753	410,598
Mymensingh	DTW	435	10,421	42,361
	STW	101,784	183,257	1,102,599
	LLP	10,398	82,810	215,665
Chattogram	DTW	82	1,930	8,997
	STW	43,957	93,175	402,389
	LLP	39,109	222,774	929,069
Sylhet	DTW	24	592	1,610
	STW	19,156	60,915	237,424
	LLP	41,853	225,867	482,919
Barishal	DTW	0	0	0
	STW	25	85	589
	LLP	25,001	156,771	380,634

DTW = deep tube well, ha = hectare, LLP = low-lift pump, STW = shallow tube well.

Source: Data extracted and adapted from Bangladesh Agricultural Development Corporation. 2020. *Minor Irrigation Survey 2020–2021*. Dhaka.

The National Minor Irrigation Policy 2017, the Groundwater Management Act 2018, and the Groundwater Management Rules 2019 clearly indicate the need for more sustainable water management in Bangladesh. Sustainable water management in irrigation requires that priority be given to the use of surface water, along with a much more efficient use of groundwater. The replacement of diesel pumps by SIP systems should follow this policy direction.

Replacing all diesel pumps with modern and efficient SIP systems would result in a total installed solar panel generating capacity of 4.4 GWp, displacing the consumption of 1 million tons of diesel fuel annually and avoiding 3 million $MtCO_2e$ every year.

Unfortunately, not all the theoretical potential can be achieved in an economical way. Economic feasibility is much lower in areas where one crop per year is the practice or in hilly areas where the installation of SIP systems will significantly be more expensive, so returns on investments will be much lower. Areas with only one crop per year usually correspond to areas prone to significant annual flooding, or areas with high-salinity soils that combine crops with other uses such as shrimp farming, which does not require irrigation.

3.3   Spatial Distribution of Irrigation Requirements

Contemporary literature on water management distinguishes water use from water consumption. The use of water retains water in liquid form but likely changes its quality. Consumption, on the other hand, transforms water from its liquid to a vapor state. In agronomy, water used for land preparation remains in liquid form, partly stored in the root zone, and the rest percolates and recharges the aquifer. Water held in the rootzone may evaporate (consumed) due to bare soil evaporation or crop transpiration.

Farmers are generally very efficient at providing water for rice when comparing the actual amount of water supplied to the field with the estimated requirements. The irrigation water delivered to the field currently averages 1,402 L per kg (kg), with a range of 951–1,671 L/kg. But not all of the water that is provided to the field is used by the plants. The true amount of water used is crop evapotranspiration, and 506–754 L are needed to grow 1 kg of rice. Percolation and seepage water return to the underlying aquifer as return flow.[40]

Irrigation requirements vary due to variations in planting dates. For Boro rice, transplanting from January onward would be the most suitable, which has been the practice in some areas, particularly in northwest Bangladesh. Due to climate change, net irrigation requirements of Boro rice may increase by a maximum of about 3% for 2050 dry climate conditions (footnote 42).

It is deduced that an increase in demand for irrigation water may come from an increase in reference evapotranspiration, an increase in irrigation area, and an increase in cropping intensity. About 95% of the total irrigation demand is for Boro rice. This is not expected to increase much from the current condition, as there has been no growth in the Boro cultivation area in recent years. There is no appreciable impact of climate change on irrigation water demand.

The Food and Agriculture Organization of the United Nations has provided 17 climate stations in Bangladesh, with additional 23 stations near its borders. These 40 stations were used to develop the crop water requirements (CWR) for Boro rice transplanted in the second half of January. The effective rainfall (Figure 2) in each *upazila* was deducted from crop evapotranspiration to arrive at the CWR. The spatial distribution of irrigation requirements is presented in Figure 4.

Measures that will help to achieve a more efficient use of water for irrigation and reduce the current dependence on diesel fuel are as follows:

- Replacing LLPs that are mostly spread across remote areas of Bangladesh with SIP systems. SIP systems could also provide electricity to some areas where no electricity supply exists. These SIP systems could offer other services, such as clean drinking water.
- Replacing STW pumps with SIP systems in sufficient number and capacities to ensure an equivalent daily discharge of water to meet the irrigation requirements of farmers.
- Replacing DTW pumps with SIP systems to provide an equivalent daily discharge. About 2,200 diesel DTWs operated by BADC could be replaced with SIP systems.

All SIP systems could export their surplus of electricity to the grid during the offseason.

---

40   M. Mainuddin et al. 2014. Bangladesh Integrated Water Resources Assessment Supplementary Report: Land Use, Crop Production, and Irrigation Demand. *Water for a Healthy Country Flagship Series*. Canberra: CSIRO.

Figure 4: Distribution of Net Boro Irrigation Requirements
(mm/Boro Season)

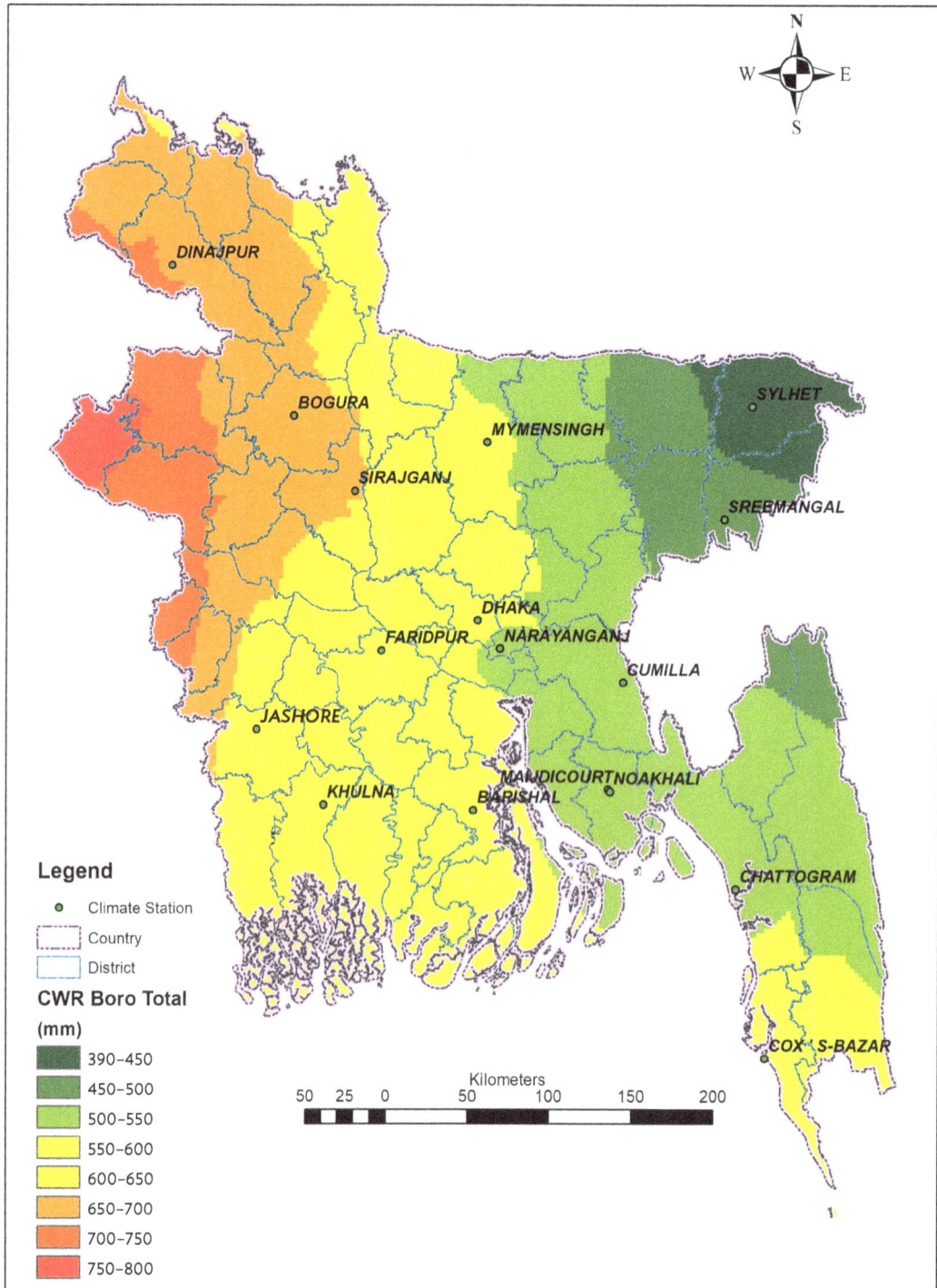

CWR = crop water requirements, mm = millimeters.

Source: Asian Development Bank consultants.

4. Available Irrigation Sources

4.1 Surface Water

Bangladesh has abundant surface water (rivers, lakes, and ponds) and groundwater resources. Rainfall and trans-boundary river flows are the main sources of surface water. Bangladesh has about 700 rivers, including tributaries and distributaries, which create about 98,000 ha of inland water bodies and more than 24,000 kilometers of streams or water channels (Figure 5). The country has an average annual surface flow of about 1.07 billion acre-feet,[41] of which about 93% is received from India as inflow (54 rivers, including the Ganges and the Brahmaputra, originate from India), and the rest as rainfall. This water is enough to cover the entire country to a depth of 9.14 m. Floods in Bangladesh cover almost one-third of its land area every year, and almost half of the country experiences severe floods. In general, seasonal flooding is shallow in the northwest, west, and east, but deep in the center and northeast. However, rivers and canals often dry up in the months of March and April, when surface water also becomes scarce. The development of irrigation with surface water is constrained due to lack of storage possibilities.

The long-term average of total renewable water resources[42] in Bangladesh is 1,211 billion cubic meters per year.[43] Bangladesh's National Water Policy of 1999 aims to increase the use of surface water as much as possible. A number of dams, barrages, and canals have been built by the Bangladesh Water Development Board (BWDB) to lessen the damaging effects of floods and utilize excess water for irrigation. The road map suggests SIP rollout for surface irrigation only in *upazilas* where surface water availability remains perennial, including in *upazilas* hosting major irrigation projects and where rubber dams are in place. The BWDB has already implemented projects to cover about 1.7 million ha of land under surface water irrigation. Village ponds and other upazilas with transient storage facilities are not included.

4.1.1 Major Existing Projects

A summary of surface irrigation projects implemented under the BWDB is presented in Table 4. The Muhuri, Ganga-Kobotakh, Chandpur, Meghna-Dhonagoda, and Pabna irrigation projects lift irrigation water using pumps from rivers or tributaries. In many instances, water is lifted by an LLP and distributed by gravity canals. These pumps have large electricity demand—the Ganges-Kobadak Irrigation Project, for instance, requires 14 MW. All of these projects offer opportunities for SIP rollout. The availability of surface water in these projects should be determined in consultation with the BWDB.

---

[41] 1 acre-foot is equivalent to 1,233.48 cubic meters.

[42] The total amount of water resources that are renewable is equal to the sum of the water resources that are obtained externally (the total amount of surface and groundwater inflows from India) and internally (the total amount of water resources that are available through endogenous precipitation).

[43] Food and Agriculture Organization of the United Nations. 2012. *Irrigation in Southern and Eastern Asia in Figures: AQUASTAT Survey 2011.* Rome.

Figure 5: Rivers of Bangladesh—Main Source of Surface Water

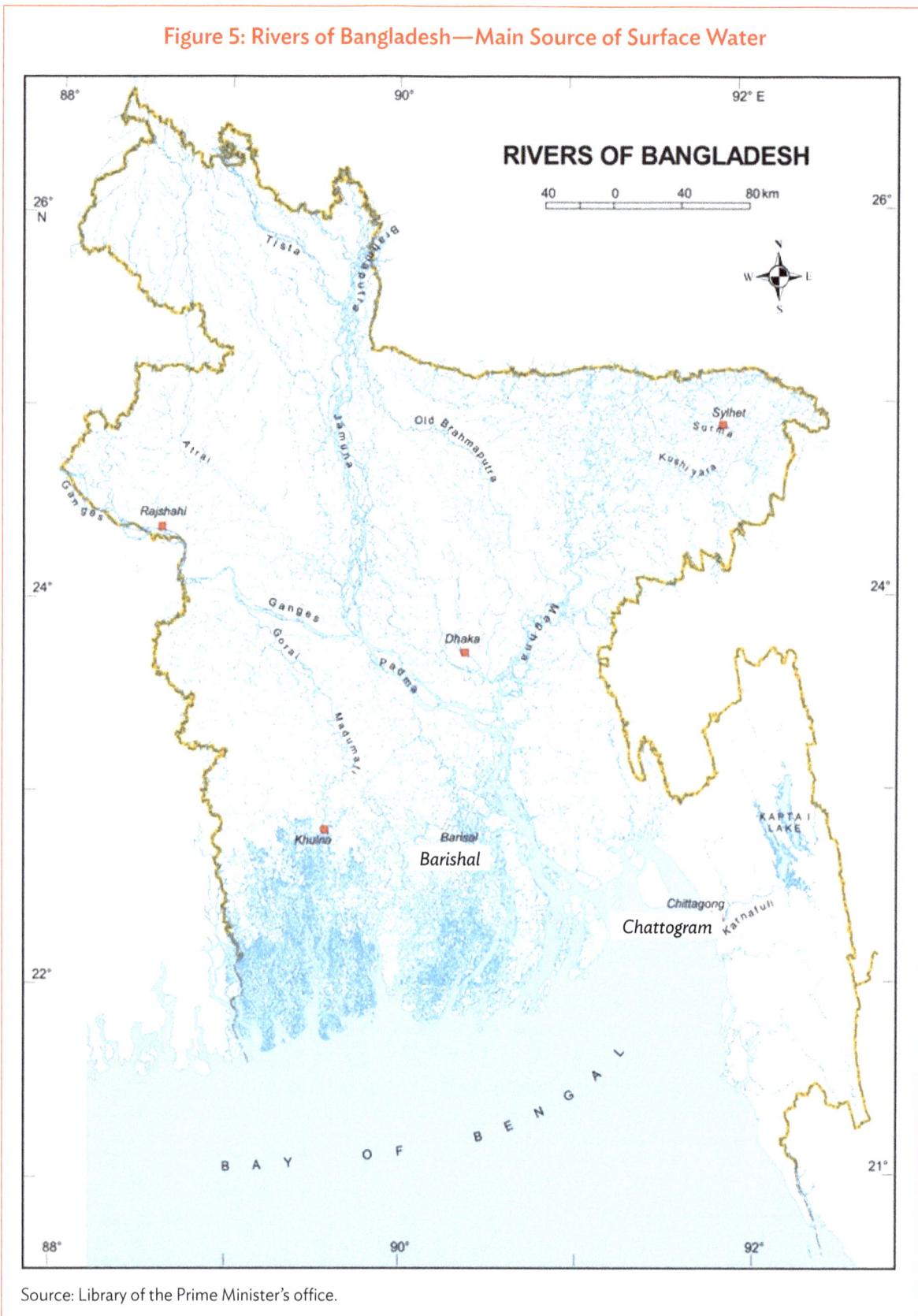

Source: Library of the Prime Minister's office.

Table 4: Summary of Major Irrigation Projects in Bangladesh

Irrigation Project Name	Location	Area (ha)
Muhuri	Feni, Sonagazi, Mirsarai, Fulgazi, and Chhagalnaiya	17,000
Ganges–Kobadak	Kushtia, Chuadanga, Jhenaidaha, and Magura	197,486
Chandpur	Chandpur and Lakshmipur	54,036
Meghna–Dhonagoda	Chandpur	19,021
Manu River	Moulvibazar	22,589
Pabna	Pabna (northwest)	196,680
Gumti	Cumilla (southeast)	37,440
Teesta Barrage	Rangpur, Lalmonirhat, and Nilphamari	132,000

ha = hectare.

Source: Asian Development Bank consultants.

4.1.2   Rubber Dams

In order to preserve water in the channels of small and medium-sized rivers, the LGED installed rubber dams to Bangladesh in 1995. This increased the amount of surface water available for irrigation and helped replenish groundwater. Irrigated areas and agricultural output increased as a result. Since then, the LGED has implemented 48 rubber dam projects for similar purposes, while the BADC has built six, BWDB five, and BMDA one.

4.1.3   Village Ponds

Over 151,000 hectares, 1.3 million ponds are found in Bangladesh. The majority of ponds ranges in size from 0.02 to 20 ha, with an average of 0.30 ha. District Barishal has the most ponds (12.11%), followed by Cumilla (9.36%), Sylhet (9.10%), Chattogram (8.02%), and Noakhali (7.75%).[44] Before tube wells were installed in rural areas in the 1960s and 1970s, ponds provided a vital source of drinking water. Today 80% of villagers use water from ponds for cooking and bathing. The local commissioner provides fishing rights. These ponds are typically small, and farmers near the ponds use the water for irrigation, usually via diesel pumps. They may be considered as a part of the SIP rollout for individuals.

4.2   Groundwater

Consultation with the BWDB should be made mandatory before the selection of aquifers for groundwater irrigation. Groundwater consists largely of surface water that has seeped down. Groundwater in Bangladesh occurs at a very shallow depth, where recent river-borne sediments form prolific aquifers in the floodplains. The groundwater table over most of Bangladesh lies close to the surface and fluctuates with the annual recharge–discharge conditions. The main component of discharge is the withdrawal of groundwater by different types of tube wells. Recharge to aquifers in Bangladesh is mainly from vertical percolation from rainwater and floodwater. A depth to groundwater level map is presented in Figure 6.

---

[44]   Government of Bangladesh, Bangladesh Bureau of Statistics. 2020. *Statistical Yearbook Bangladesh 2019*. 39th Edition. Dhaka

Figure 6: Groundwater Zoning Map, 2020

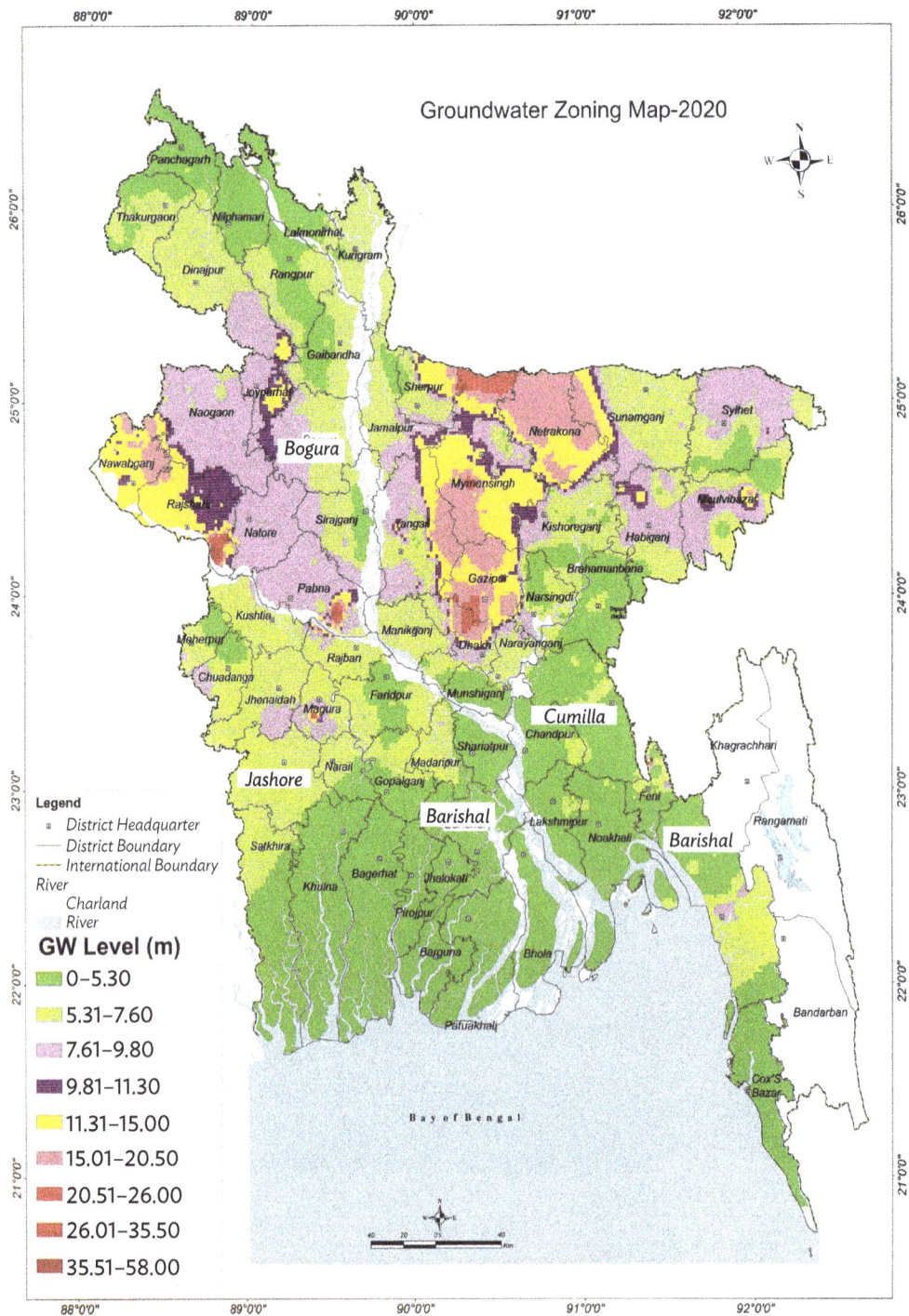

GW = groundwater, m = meters.

Source: Bangladesh Agricultural Development Corporation. 2020. *Minor Irrigation Survey Report 2020–2021 (Figure 39, page 26)*. Dhaka.

4.2.1   Hydrogeology and River-Aquifer Interaction

At the outlet of the Bay of Bengal, Bangladesh covers a significant part of the Bengal Delta, which was created by the deposition of the Ganges, Brahmaputra, and Meghna river systems. The unconsolidated near-surface Pleistocene to Recent fluvial and estuarine sediments underlying most of Bangladesh generally form prolific aquifers. In central Bangladesh, an aquitard delineates the boundaries of the shallow and deep aquifers at a depth of 150 meters; however, the boundary extends deep in the southern coastal districts.

In 1975, an innovative solution to freshwater storage in the River Ganges Basin was implemented, which saw incremental increases in dry-season groundwater pumping for irrigation near river channels.[45] Dubbed the "Ganges Water Machine," this intervention sought to increase the capture and storage of seasonal freshwater surpluses while mitigating monsoonal flood risk.

Others described this broader set of recharge pathways caused by dry-season groundwater pumping the "Bengal Water Machine" (BWM) and expanded the idea of freshwater capture of monsoonal flows beyond perennial rivers to include a range of surface waters, such as ponds, canals, and seasonal rivers. They also enhanced local drainage and diffused recharge through irrigation return flows in the Bengal Basin.[46] In the Bengal Basin of Bangladesh, there is a strong correlation between river discharge and groundwater levels.[47] The proven resistance of this combined use of surface water and groundwater to hydrological extremes that are exacerbated by climate change is of strategic significance.

Locations where monsoon-induced recharge is insufficient to fully replenish groundwater abstracted during the dry season are among the major BWM operation limitations. For instance, regions with low permeability surface geology limit the BWM and correspond with groundwater levels over 8 m below ground during the dry season, making groundwater unusable for homes that depend on shallow wells operated by suction. Moreover, the highest risk of reaching the limits of enhanced freshwater capture through the BWM is found in the High Barind region and Ganges floodplain in western Bangladesh, where observed groundwater recharge approaches or surpasses potential recharge and is determined by rainfall, surface geology, and flood extent. As such, chances to extend the BWM's operations in Bangladesh are currently primarily limited to the floodplains of the Brahmaputra River.

4.2.2   Fluctuations and Decline

Groundwater levels in heavily irrigated areas vary from 5 to 15 meters below the surface, and in certain locations, during the dry peak irrigation season, they can reach up to 23 meters below the surface (excluding the Dhaka city area).[48]

45   R. Revelle and V. Lakshminarayana. 1975. The Ganges Water Machine. Science. 188 (4188). pp. 611–616. https://www.science.org/doi/10.1126/science.188.4188.611.

46   M. Shamsudduha et al. 2022. The Bengal Water Machine: Quantified Freshwater Capture in Bangladesh. Science. 377 (6612). pp. 1315–1319.

47   A. Zahid et al. 2009. The Impact of Shallow Tubewells on Irrigation Water Availability, Access, Crop Productivity and Farmers' Income in the Lower Gangetic Plain of Bangladesh. In A. Mukherji et al. (eds). Groundwater Governance in the Indo-Gangetic and Yellow River Basins: Realities and Challenges. London: CRC Press.

48   Government of Bangladesh, General Economic Division—Bangladesh Planning Commission, Ministry of Planning. 2018. *Bangladesh Delta Plan 2100: Bangladesh in the 21st Century*. Dhaka. https://oldweb.lged.gov.bd/UploadedDocument/UnitPublication/1/756/BDP%202100%20Abridged%20Version%20English.pdf.

Weekly monitoring records of groundwater levels throughout Bangladesh showed that shallow groundwater levels have been declining at a high rate recently (i.e., 1985–2005).[49] Declining rates are highest (exceeding −0.5 m per year) in and around Dhaka city and the High Barind Tract region and high (0 to −0.05 m per year) in areas south of the Ganges River. In coastal areas, shallow groundwater levels show stable to slightly rising trends (0 to +0.1 m per year) over the same period.

Several other studies have examined fluctuations, including the following:

- An investigation over a period of 25 years comprehensively analyzed changes to groundwater levels in Barind and noted that as the number of DTWs and the quantity of water abstracted increased, the magnitude of the fluctuations became larger.[50] These studies observed variation in the recovery rate among the three landforms, High Barind, Level Barind, and Flooded Terrace, at Barind Tract.
- The Bangladesh Rural Advancement Committee discovered a 30-year trend of declining groundwater levels in the northwest (1981–2011).[51] Rajshahi was the district most severely affected, followed by Pabna, Bogura, Dinajpur, and Rangpur.
- Groundwater level decline has also been observed in Bogura district, northwest Bangladesh, where open-cut coal mining—Barapukuria Coal Mine in Bogura district is the only active coal mine in the country—and sand mining take place.[52] Together, both activities led to groundwater levels falling due to an increase in void ratio. Consequently, many water bodies near the mine have dried up, probably due to water flowing into the mine pits, possibly forming permanent lakes. This requires further investigation. Surface subsidence was first observed in 2006, evident from cracks in the surface structures of the mining area, and the government has acquired 2.61 square kilometers of the affected land area as a form of compensation because land was no longer productive for the farmers (footnote 53).
- The BWDB meanwhile analyzed hydrographs from 211 representative observation wells from 2008 to 2018 in eastern Bangladesh.[53] The study area boundary from north to south was lined by Jamuna, Padma, and Meghna rivers; this area covers the Brahmaputra-Jamuna floodplain, old Brahmaputra floodplain, Madhupur tract, Sylhet depression, Meghna floodplain, Tippera surface, Chattogram Hill tracts, and some parts of Ganges delta plain. Most of the study area wells showed stable trends with inevitable seasonal fluctuations, which means the groundwater levels in those areas were stable and in good condition/amounts.

4.2.3   Hazards Mapping

In a 2020 hazards study, a multi-parameter groundwater hazard map of Bangladesh was developed, combining information on arsenic, salinity, and water storage.[54] In order to estimate exposures, a variety of socioeconomic variables were also superimposed on these risk maps, such as social vulnerability (i.e., poverty) and access to drinking and irrigation water supplies. The resulting maps demonstrate that the combined threat of groundwater storage depletion and contamination by arsenic and salinity currently affects a significant portion of land area

49   M. Shamsudduha et al. 2011. The Impact of Intensive Groundwater Abstraction on Recharge to a Shallow Regional Aquifer System: Evidence from Bangladesh. *Hydrogeology Journal*. 19. pp. 901–916. DOI 10.1007/s10040-011-0723-4.

50   K. Rushton. 2022. *Understanding Groundwater Resources in the Rajshahi Barind: An Investigation Over a Period of 25 Years*. Birmingham: Birmingham University.

51   Bangladesh Rural Advancement Committee. 2013. *Sustainability of Groundwater Use for Irrigation in Northwest Bangladesh*. Dhaka.

52   A.K.M.B. Alam et al. 2022. Prediction of Mining-Induced Subsidence at Barapukuria Longwall Coal Mine, Bangladesh. *Scientific Reports*. 12 (14800). https://doi.org/10.1038/s41598-022-19160-1.

53   BWDB. 2020. Groundwater Table Hydrograph of 38 Districts for the Year 2008 to 2018. Dhaka. http://www.hydrology.bwdb.gov.bd/img_upload/ongoing_project/781.pdf.

54   M. Shamsudduha et al. 2019. Multi-Hazard Groundwater Risks to Water Supply from Shallow Depths: Challenges to Achieving the Sustainable Development Goals in Bangladesh. *Exposure and Health*. 12. pp. 657–670. https://doi.org/10.1007/s12403-019-00325-9.

(5%–24% under high to extremely high risks). This means effectively 6.5 million people (of whom 2.2 million are poor) to 24.4 million people (of whom 8.6 million are poor) are exposed to the combined risks of high arsenic, salinity, and groundwater storage depletion.

Deep groundwater (beyond 150 meters below ground level) in the Bengal Aquifer System is extensively utilized for domestic water supply throughout the floodplains of southern Bangladesh. This is a de facto mitigation measure against the high levels of arsenic in shallow groundwater, which surpasses the World Health Organization standard of 10 µg/L. Studies show that deep groundwater will remain secure against the invasion of arsenic across the entire region if it is restricted to domestic use, even under domestic demand projected for 2050.[55] Hence, groundwater for irrigation should not be pumped at depths of more than 150 m.

4.3   Anomalies in Terrestrial Water Storage

The study used the Gravity Recovery and Climate Experiment (GRACE) and Follow On (GRACE FO) mission data to compute the spatiotemporal variations in groundwater storage (GWS) and terrestrial water storage (TWS) across Bangladesh. Anomalies observed in pre- (May) and post- (Nov) monsoon were analyzed to determine whether the changes between 2002 and 2022 were structural and statistically significant, or fluctuations due to random factors.

Out of the 495 *upazilas*, the maximum increase (0.44 cm per year) in pre-monsoon TWS was in Lohagara, and the minimum change (–3.17 cm per year) was in Bhurungamari. About 195 *upazilas* showed a decline in TWS, of which changes in 98 were statistically significant. About 300 *upazilas* showed an increase in TWS, of which the rise in 10 was significant.

Out of the 495 *upazilas*, the maximum increase (0.66 cm per year) in post-monsoon anomaly was in Bera, and the minimum change (–3.17 cm per year) was in Bhurungamari. About 170 *upazilas* showed a decline in TWS, of which 63 was statistically significant. About 325 *upazilas* showed an increase in TWS, of which the rise in 120 was significant.

From the above, it can be concluded that water consumption is increasing in 195 *upazilas*, significantly in 98 of them. However, substantial recovery in water storage occurs during monsoon, resulting in only 63 *upazilas* where TWS declines annually for a maximum of 3.7 cm per year. SIP rollout in these 63 *upazilas* should be carefully assessed for their water use and is recommended to be accompanied by water conservation measures.

Out of the 495 *upazilas*, the maximum increase in GWS (1.2 cm per year) in pre-monsoon anomaly was in Ramgarh, and the minimum change of –2.35 cm per year was in Domar. About 170 *upazilas* showed a decline in GWS, of which changes in 135 were statistically significant. About 325 *upazilas* showed an increase in GWS, of which the rise in 68 *upazilas* was significant.

---

[55] M. Shamsudduha, A. Zahid, and W. G. Burgess. 2019. Security of Deep Groundwater against Arsenic Contamination in the Bengal Aquifer System: A Numerical Modeling Study in Southeast Bangladesh. *Sustainable Water Resources Management*. 5. pp. 1073–1087. https://doi.org/10.1007/s40899-018-0275-z.

Out of the 495 *upazilas*, the maximum increase of GWS (1.02 cm per year) in post-monsoon anomaly was in Sreepur, and the minimum change (–2.6 cm per year) was in Domar. About 172 *upazilas* showed a decline in GWS, of which 37 were statistically significant changes. About 323 *upazilas* showed an increase in GWS, of which the rise in 239 *upazilas* was significant.[56]

From the above, it may be concluded that groundwater consumption is increasing in 170 *upazilas*, significantly in 68. However, substantial recovery in GWS occurs during the monsoon, resulting in only 37 *upazilas* where GWS declines annually by a maximum of 2.7 cm per year. SIP rollout to pump groundwater in these 37 *upazilas* may be carried out, with water conservation measures such as managing aquifer recharges implemented, to reduce the rate of decline.

This overall analysis shows that TWS (Figure 7) is declining steadily in 63 *upazilas* and GWS is declining steadily in 37 *upazilas* (Figure 8). It is noted that negative anomalies indicate that groundwater storage is declining and positive anomalies indicate that groundwater is increasing. The maximum rate of decline of TWS is –3.17 cm per year and GWS is –2.6 cm per year, which are relatively small. However, the decline has been steady in these *upazilas*. Hence, water conservation measures to reduce water consumption (such as alternate wetting and drying in crop production) and increase storage (managed aquifer recharge) must be adopted to overcome these problems.

---

[56]   In these *upazilas*, new SIPs (in addition to replacement of existing wells) may be implemented, provided they are not affected by salinity, flood, tide, or arsenic.

Figure 7: Changes to Post-Monsoon Terrestrial Water Storage Estimated from Gravity Recovery and Climate Experiment Anomalies
(cm/year)

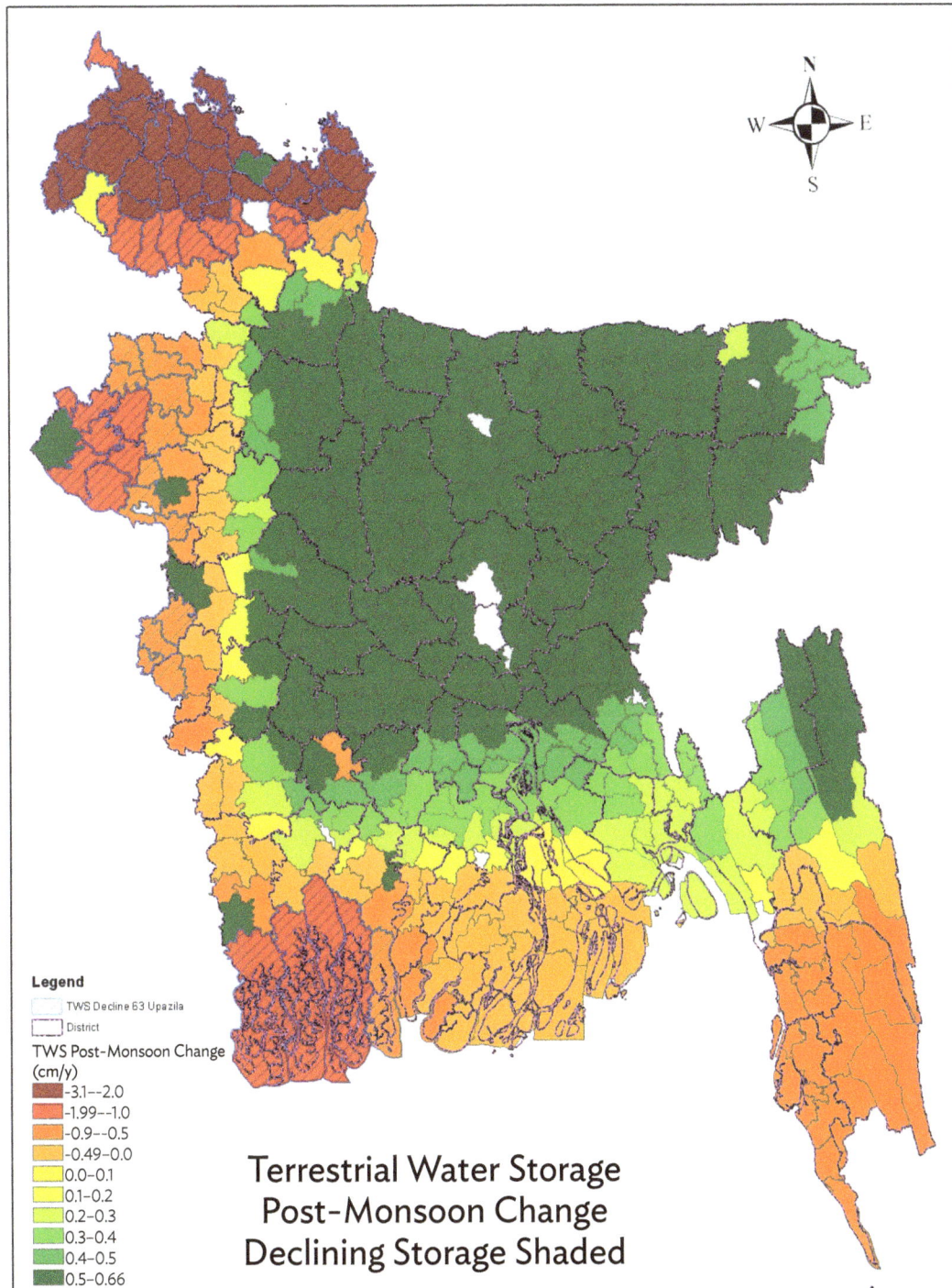

Legend

TWS Decline 63 Upazila

District

TWS Post-Monsoon Change (cm/y)
- -3.1–-2.0
- -1.99–-1.0
- -0.9–-0.5
- -0.49–-0.0
- 0.0–0.1
- 0.1–0.2
- 0.2–0.3
- 0.3–0.4
- 0.4–0.5
- 0.5–0.66

Terrestrial Water Storage Post-Monsoon Change Declining Storage Shaded

cm = centimeter, TWS = terrestrial water storage

Source: Asian Development Bank consultants.

Figure 8: Changes to Post-Monsoon Groundwater Storage Estimated from Gravity Recovery and Climate Experiment Anomalies
(cm/year)

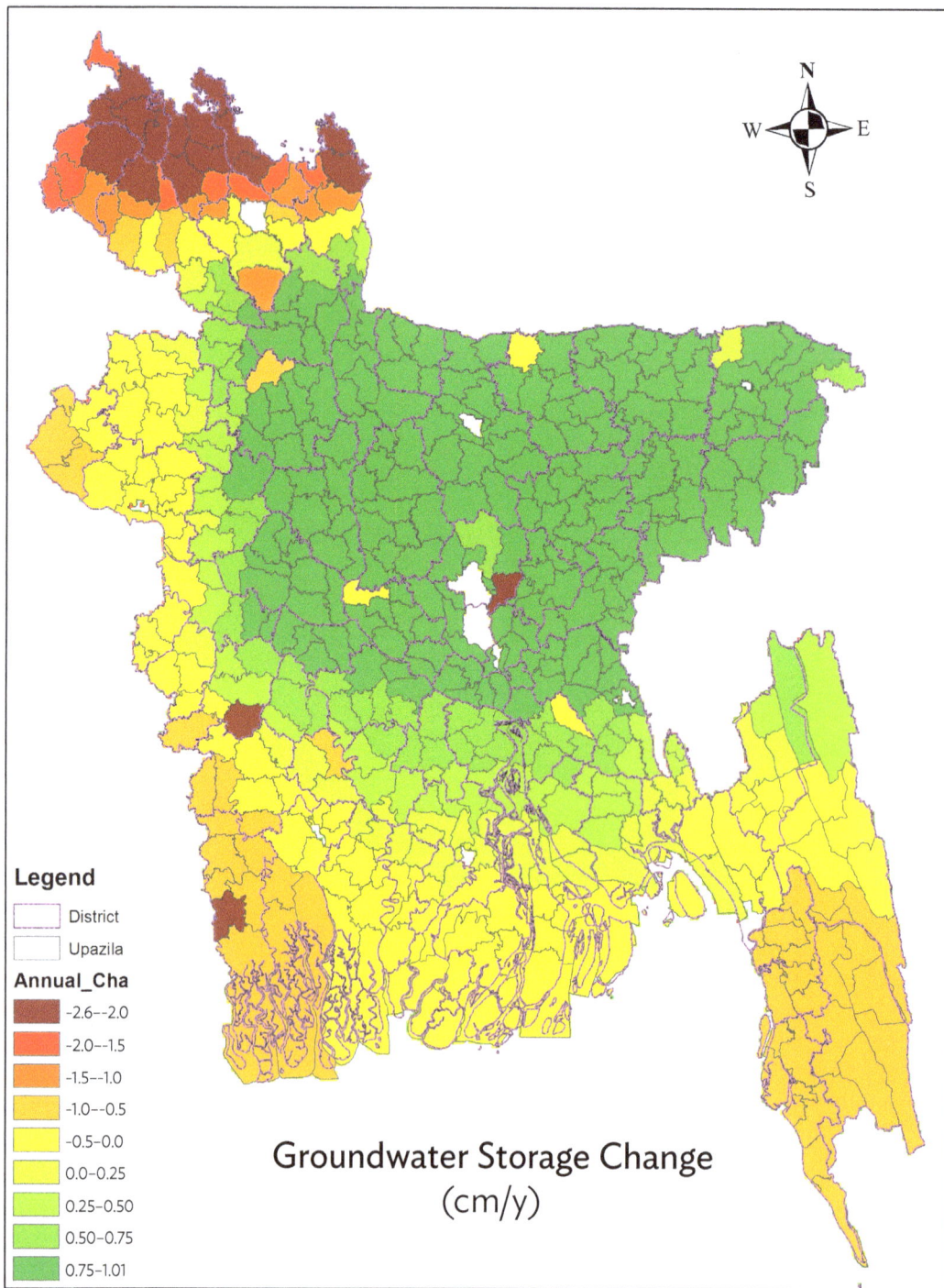

Source: Asian Development Bank consultants.

5. Multi-Criteria Assessment for the Selection of Suitable Areas for SIP Rollout

5.1 High Potential Locations for Surface Water Irrigation

Opportunities for SIP rollout within large surface irrigation schemes appear limited, although existing electrified LLPs may be replaced with SIP systems. In most instances, water flows from dams due to gravity; hence, the need to lift water further does not arise. Circumstances may also cause canals to run dry, likely at the tail ends of the system. Rubber dams, on the other hand, provide the potential to develop small (200 ha) irrigation schemes. The *upazilas* identified for surface water irritation supported with SIP are shown in Figure 9.

Figure 9: *Upazilas* with Rubber Dams Selected for Surface SIPs

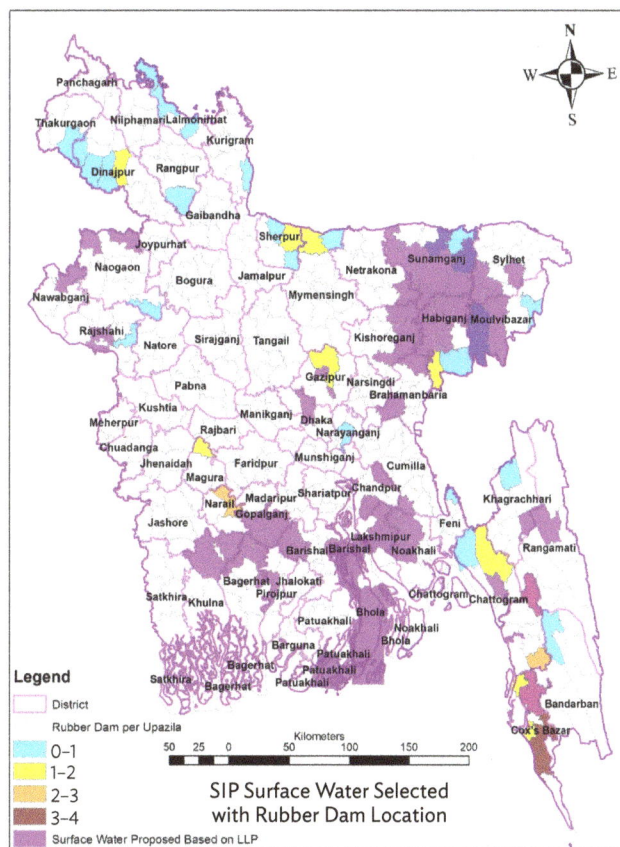

LLP = low-lift pump, SIP = solar irrigation pump.
Source: Asian Development Bank consultants.

There are 75 *upazilas* recommended for surface water irrigation under the road map (Table 5). It is important to mention that inclusion or non-inclusion does not mean a particular SIP can or cannot be installed in these locations. Each individual location must be taken on its merits according to prevailing criteria at that site.

Table 5: Selected *Upazilas* for Surface Water Solar Irrigation Pumps Rollout

Division	District	*Upazila*	District	*Upazila*	District	*Upazila*
Barishal	Barishal	Agailjhara	Bhola	Bhola Sadar	Bhola	Tazumuddin
	Barishal	Banaripara	Bhola	Borhanuddin	Patuakhali	Rangabali
	Barishal	Mehendiganj	Bhola	Charfasson	Pirojpur	Indurkani
	Barishal	Ujirpur	Bhola	Lalmohan	Pirojpur	Pirojpur Sadar
Chattogram	Brahmanbaria	Nabinag'r	Cox's Bazar	Chakaria	Noakhali	Chatkhil
	Brahmanbaria	Nasirnagar	Cumilla	Manoharganj	Noakhali	Noakhali Sadar
	Chandpur	Faridganj	Khagrachhari	Mahalchhari	Noakhali	Sonaimuri
	Chandpur	Kachua	Lakshmipur	Lakshmipur Sadar	Rangamati	Langadu
	Chattogram	Hathazari	Lakshmipur	Ramganj		
	Chattogram	Rangunia	Noakhali	Begumganj		
Dhaka	Dhaka	Savar	Gopalganj	Kotalipara	Kishoreganj	Bajitpur
	Gazipur	Gazipur Sadar	Gopalganj	Tungipara	Kishoreganj	Itna
	Gopalganj	Gopalganj Sadar	Kishoreganj	Austagram	Kishoreganj	Mithamain
Khulna	Bagerhat	Chitalmari	Khulna	Batiaghata	Khulna	Rupsa
	Bagerhat	Fakirhat	Khulna	Dighalia	Khulna	Terokhada
	Bagerhat	Kachua	Khulna	Dumuria		
	Bagerhat	Mollahat	Khulna	Phultala		
Mymensingh	Netrakona	Khaliajuri				
Rajshahi	Chapainababganj	Gomastapur	Naogaon	Sapahar		
	Naogaon	Dhamoirhat	Rajshahi	Paba		
Sylhet	Habiganj	Ajmiriganj	Moulvibazar	Rajnagar	Sunamganj	Shalla
	Habiganj	Baniachong	Moulvibazar	Sreemangal	Sunamganj	Shantiganj
	Habiganj	Habiganj Sadar	Sunamganj	Chhatak	Sunamganj	Sunamganj Sadar
	Habiganj	Lakhai	Sunamganj	Derai	Sylhet	Balaganj
	Habiganj	Nabiganj	Sunamganj	Dharmapasha	Sylhet	Bishwanath
	Habiganj	Shayestaganj	Sunamganj	Jagannathpur	Sylhet	Golapganj
	Moulvibazar	Kamalganj	Sunamganj	Jamalganj	Sylhet	Osmaninagar
	Moulvibazar	Moulvibazar Sadar	Sunamganj	Madhyanagar		

Source: Asian Development Bank consultants.

5.2   Selection of Suitable Locations for Groundwater Irrigation

A geographic information system-based multi-criteria analysis (MCA) was conducted by assigning scores between 0 and 1 to eight factors that will affect the success of the SIP rollout. Simulations were not done to determine the sensitivity of weights used in the MCA. Weights were assigned to the seven factors based on available public data (soil type had no available data, so it was not included). A summary of the factors, scores, and justification for the scores is presented in Table 6. A total score for each *upazila* was determined by adding the seven factor scores. *Upazilas* with scores above 3.5 are considered most favorable for SIP rollout for groundwater irrigation (locations with potential level 1) (Figure 10). *Upazilas* with a total score between 2.5 and 3.5 were recommended for SIP rollout with the adoption of water conservation measures (locations with potential level 2) (Figure 11). *Upazilas* with a total score of less than 2.5 are recommended for SIP rollout with caution and only after detailed analysis of the seven factors in the proposed sites (locations with potential level 3). However, it must be emphasized that consultation with the BWDB should be made mandatory before the selection of aquifers for groundwater irrigation.

Table 6: Weights Assigned to Different Ranges of Each Factor Considered in the Multi-Criteria Analysis

No	Factor	Range	Weight	Justification
1	Salinity (PS/cm)	<700	1	Most suitable water for irrigation
		700–3000	0.5	May be used for irrigation
		>3000	0	Unsuitable for irrigation
2	Arsenic (µg/L)	<10	1.0	
		10–50	0.5	
		>50	0	
3	Flood Depth[a] (cm)	<90	1	Minor hazard
		90–150	0.25–0.75	Moderate hazard
		151–240	0–0.25	High hazard
		>240	0	Very high hazard
4	Tidal Surge	Yes	0	
		No	1	
5	Tube Well Spacing in Meters for Pumps of 0.5 Cusecs Discharge	0	0	Surrogate for groundwater withdrawal as a fraction of net recharge
		0–240	1	
		240–350	0.9	
		>350	0.5	Provides more weight to undeveloped recharge potential
6	Groundwater Storage Change (cm/y)	–2.6– –0.5	0.1	Changes to groundwater stored—negative informs discharge exceeding recharge
		–0.5–0.0	0.33	
		0.0–0.5	0.66	
		0.5–1.0	1	
7	Boro Crop Water Requirement (mm/season)	421–500	1	Lesser the requirement, the less stress on groundwater resources
		500–600	0.66	
		600–700	0.33	
		700–800	0	

cm = centimeter, m = meter, µg/L = microgram per liter, mm = millimeter, PS = potential salinity, y = year.

[a] Flood depth has been used as a proxy for hills as well as for areas with low drainage.

Source: Asian Development Bank consultants.

The caveat of the MCA:

- The MCA identified seven criteria that affect the irrigation potential of groundwater. In alphabetical order, they are as follows: arsenic, changes to groundwater stored, crop water requirements, flood depth, maximum surge level, salinity, and recommended spacing for tube wells (as a surrogate for undeveloped groundwater recharge).
- Prevailing soil type, which has a large bearing on the SIP design, was excluded as a criterion because no detailed soil map was available. Even if available, a homogeneous soil type would have to be assigned to each *upazila*, which would be an unrealistic assumption.
- Climate change impacts crop evapotranspiration, effective rainfall, and recharge to groundwater. The impact of Climate Change on Groundwater potential will need to be considered in the future.
- The MCA is based on publicly available data, mostly reported in peer-reviewed publications or reports published by government agencies.
- Weights were assigned to the seven factors that affect groundwater resource potential at any location. The range of weights was based on the significance of the factor to groundwater resource potential and the extent of variation of the factor. For example, the weights of GWS were within a range of 0.1 and 1, but the weights for flood depth were between 0 and 1. Changes to the weights and corresponding ranges of the factor may affect the final results.
- Hence, the MCA results may be considered as a preliminary assessment of the groundwater resource potential of an *upazila*. Inclusion or non-inclusion in the selection level of an *upazila* does not mean a particular SIP can or cannot be installed. Irrigation patterns vary from district to district, even in some cases from subdistrict to subdistrict. Therefore, this road map recommends designing the SIP based on prevailing conditions. Before design and implementation, each project area should be surveyed in detail.
- The MCA categorizes four levels of groundwater resource potential based on the seven criteria. Potential level 1 *upazilas* have the most favorable groundwater conditions for the rollout; potential levels 2 and 3 *upazilas* have one or more of the seven detrimental criteria. Therefore, one or more constraining criteria need to be identified at that location, and the decision to proceed should be based on acceptable critical values. Potential level 4 *upazilas* are not recommended by the government for groundwater development based on the Integrated Water Management tube well spacing report.

5.2.1   Locations with Potential Level 1: *Upazilas* Where Current Recharge Mechanisms Are Adequate to Sustain Groundwater Irrigation

The MCA identified 158 *upazilas* where current recharge mechanisms are adequate to sustain groundwater irrigation. These *upazilas* are those where the total score of the MCA was higher than 3.5. In these *upazilas*, SIP rollout is feasible with no major constraints (Figure 10 and Table 7). It is important to note that inclusion or non-inclusion in the listed *upazilas* does not mean a particular SIP can or cannot be installed in these locations. Each individual project site must be taken on its own merits according to prevailing conditions. Consultation with the BWDB should also be done before selection of aquifers for groundwater irrigation.

Figure 10: *Upazilas* Where Groundwater Solar Irrigation Pumps Are Feasible
with No Major Constraints

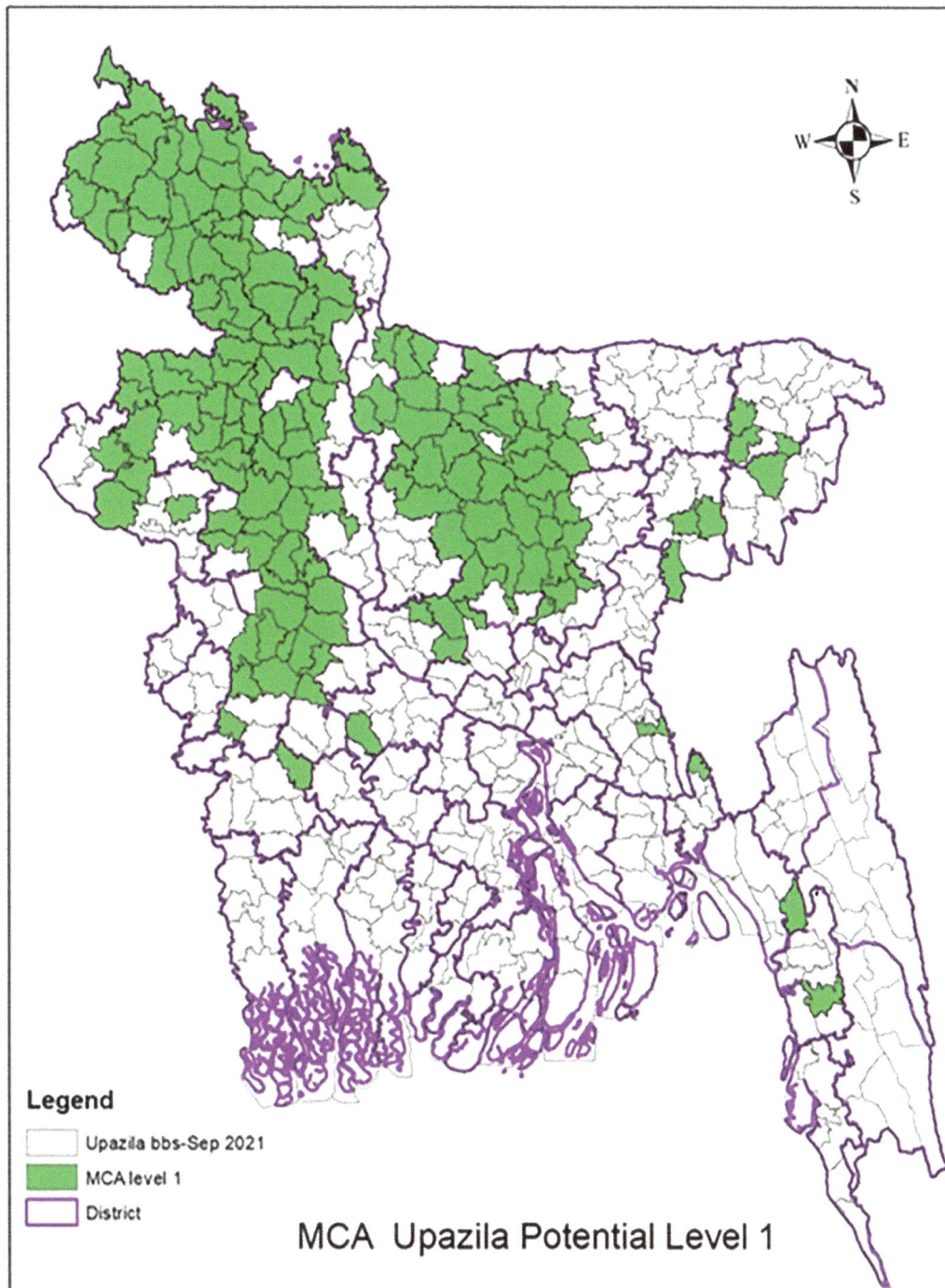

Legend

Upazila bbs-Sep 2021
MCA level 1
District

MCA Upazila Potential Level 1

bbs = Bangladesh Bureau of Statistics, MCA = multi-criteria analysis.
Source: Asian Development Bank consultants.

Table 7: Adequate *Upazilas* to Sustain Groundwater Irrigation with Solar Irrigation Pump Systems

District	Upazila	District	Upazila	District	Upazila
Bogura	Adamdighi	Joypurhat	Khetlal	Netrakona	Kendua
Bogura	Bogura Sadar	Joypurhat	Panchbibi	Netrakona	Netrakona Sadar
Bogura	Dupchachia	Kishoreganj	Hossainpur	Netrakona	Purbadhala
Bogura	Gabtali	Kishoreganj	Katiadi	Nilphamari	Dimla
Bogura	Kahaloo	Kishoreganj	Kishoreganj Sadar	Nilphamari	Domar
Bogura	Nandigram	Kishoreganj	Pakundia	Nilphamari	Jaldhaka
Bogura	Shajahanpur	Kurigram	Bhurungamari	Nilphamari	Kishoreganj
Bogura	Sherpur	Kurigram	Nageshwari	Nilphamari	Nilphamari Sadar
Bogura	Sonatala	Kurigram	Phulbari	Nilphamari	Saidpur
Chapainababganj	Nachole	Kushtia	Khoksa	Pabna	Atgharia
Chattogram	Raozan	Kushtia	Kumarkhali	Pabna	Bhangura
Chattogram	Satkania	Kushtia	Kushtia Sadar	Pabna	Chatmohar
Cumilla	Sadar Dakkhin	Lalmonirhat	Aditmari	Pabna	Faridpur
Dhaka	Dhamrai	Lalmonirhat	Hatibandha	Pabna	Pabna Sadar
Dhaka	Savar	Lalmonirhat	Kaliganj	Pabna	Santhia
Dinajpur	Birampur	Lalmonirhat	Lalmonirhat Sadar	Pabna	Sujanagar
Dinajpur	Birganj	Lalmonirhat	Patgram	Panchagarh	Atowari
Dinajpur	Bochaganj	Magura	Sreepur	Panchagarh	Boda
Dinajpur	Chirirbandar	Manikganj	Saturia	Panchagarh	Debiganj
Dinajpur	Dinajpur Sadar	Manikganj	Singair	Panchagarh	Panchagarh Sadar
Dinajpur	Fulbari	Moulvibazar	Rajnagar	Panchagarh	Tentulia
Dinajpur	Ghoraghat	Mymensingh	Bhaluka	Rajbari	Kalukhali
Dinajpur	Hakimpur	Mymensingh	Fulbaria	Rajbari	Pangsha
Dinajpur	Kaharole	Mymensingh	Fulpur	Rajshahi	Durgapur
Dinajpur	Khansama	Mymensingh	Gafargaon	Rajshahi	Godagari
Dinajpur	Nababganj	Mymensingh	Gouripur	Rajshahi	Tanore
Dinajpur	Parbatipur	Mymensingh	Haluaghat	Rangpur	Badarganj
Faridpur	Boalmari	Mymensingh	Ishwarganj	Rangpur	Gangachara
Feni	Parashuram	Mymensingh	Muktagachha	Rangpur	Kaunia
Gaibandha	Gaibandha Sadar	Mymensingh	Mymensingh Sadar	Rangpur	Mithapukur
Gaibandha	Gobindaganj	Mymensingh	Nandail	Rangpur	Pirganj
Gaibandha	Palashbari	Mymensingh	Tarakanda	Rangpur	Rangpur Sadar
Gaibandha	Sadullapur	Mymensingh	Trishal	Rangpur	Taraganj

continued on next page

Table 7 *continued*

District	Upazila	District	Upazila	District	Upazila
Gaibandha	Saghata	Naogaon	Atrai	Sherpur	Jhenaigati
Gaibandha	Sundarganj	Naogaon	Badalgachhi	Sherpur	Nakla
Gazipur	Gazipur Sadar	Naogaon	Dhamoirhat	Sherpur	Sherpur Sadar
Gazipur	Kaliakair	Naogaon	Mahadebpur	Sherpur	Sreebardi
Gazipur	Kapasia	Naogaon	Naogaon Sadar	Sirajganj	Kamarkhanda
Gazipur	Sreepur	Naogaon	Niamatpur	Sirajganj	Rayganj
Habiganj	Bahubal	Naogaon	Patnitala	Sirajganj	Tarash
Habiganj	Habiganj Sadar	Naogaon	Porsha	Sylhet	Bishwanath
Habiganj	Madhabpur	Naogaon	Raninagar	Sylhet	Fenchuganj
Jamalpur	Bakshiganj	Naogaon	Sapahar	Sylhet	Osmaninagar
Jamalpur	Jamalpur Sadar	Narsingdi	Belabo	Tangail	Dhanbari
Jamalpur	Madarganj	Narsingdi	Manohardi	Tangail	Ghatail
Jamalpur	Melandaha	Narsingdi	Palash	Tangail	Gopalpur
Jashore	Bagharpara	Narsingdi	Shibpur	Tangail	Madhupur
Jhenaidah	Harinakundu	Natore	Bagatipara	Tangail	Sakhipur
Jhenaidah	Kotchandpur	Natore	Baraigram	Thakurgaon	Baliadangi
Jhenaidah	Shailkupa	Natore	Gurudaspur	Thakurgaon	Pirganj
Joypurhat	Akkelpur	Natore	Natore Sadar	Thakurgaon	Ranishankail
Joypurhat	Joypurhat Sadar	Natore	Singra	Thakurgaon	Thakurgaon Sadar
Joypurhat	Kalai	Netrakona	Atpara		

Source: Asian Development Bank Consultants.

5.2.2   Locations with Potential Level 2: **Upazilas** Where Water Conservation Measures Are Needed to Sustain Groundwater Use

The MCA identified 137 *upazilas* where water conservation measures are needed to sustain groundwater irrigation. These *upazilas* are those where the total score of the MCA was higher than 2.5, but lower than 3.5. It is recommended that water harvesting measures are implemented in water-deficit *upazilas* when rolling out the road map. In these *upazilas,* SIP rollout is feasible when water conservation measures are implemented (Figure 11 and Table 8). It is important to monitor any decline in groundwater level in these *upazilas,* as that should be considered as part of its selection. As noted, inclusion or non-inclusion in the listed *upazilas* does not mean a particular SIP can or cannot be installed in these locations. Each individual project site must be taken on its own merits according to prevailing conditions.

Figure 11: *Upazilas* **Where Groundwater Solar Irrigation Pumps Are Feasible and a Rollout Is Recommended if Water Conservation Measures Are Implemented**

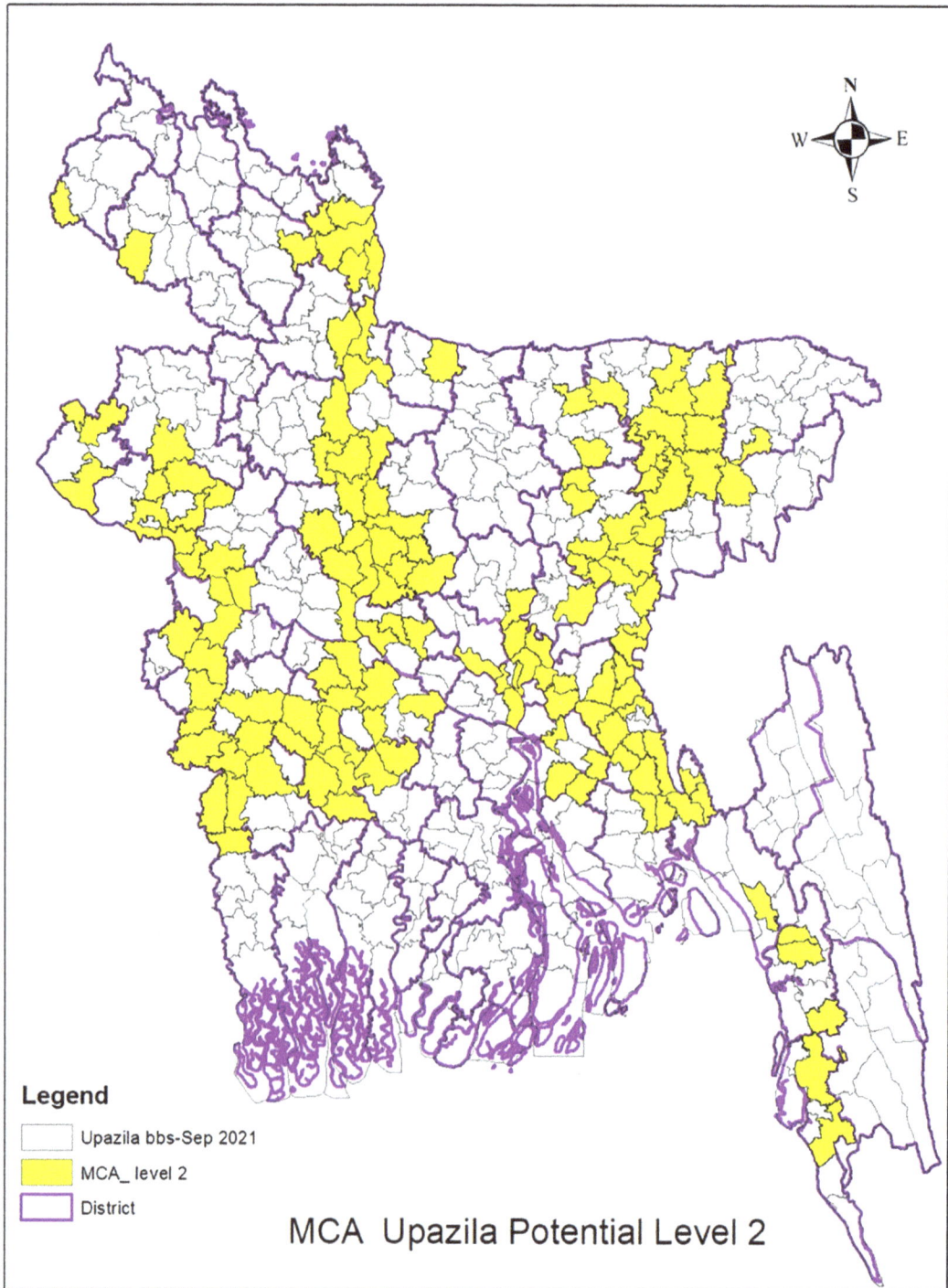

Legend
Upazila bbs-Sep 2021
MCA_ level 2
District

MCA Upazila Potential Level 2

bbs = Bangladesh Bureau of Statistics, MCA = multi-criteria analysis.
Source: Asian Development Bank consultants.

Table 8: Selected *Upazilas* for Groundwater SIP Rollout on Condition that Water Conservation Measures Are Implemented

District	*Upazila*	District	*Upazila*	District	*Upazila*
Bogura	Dhunat	Gaibandha	Fulchhari	Narayanganj	Bandar
Bogura	Sariakandi	Gazipur	Kaliganj	Narayanganj	Narayanganj Sadar
Brahmanbaria	Akhaura	Gopalganj	Kashiani	Narayanganj	Rupganj
Brahmanbaria	Bijoynagar	Gopalganj	Muksudpur	Narayanganj	Sonargaon
Brahmanbaria	Kasba	Habiganj	Ajmiriganj	Narsingdi	Raipura
Brahmanbaria	Nasirnagar	Habiganj	Baniachong	Natore	Lalpur
Brahmanbaria	Sarail	Habiganj	Lakhai	Natore	Naldanga
Chandpur	Faridganj	Habiganj	Nabiganj	Netrakona	Barhatta
Chandpur	Matlab Dakkhin	Jamalpur	Dewanganj	Netrakona	Madan
Chandpur	Shahrasti	Jamalpur	Islampur	Noakhali	Senbag
Chapainababganj	Bholahat	Jamalpur	Sarishabari	Pabna	Bera
Chapainababganj	Chapainawabganj Sadar	Jashore	Chaugachha	Pabna	Ishwardi
Chapainababganj	Gomastapur	Jashore	Jashore Sadar	Rajbari	Baliakandi
Chattogram	Boalkhali	Jashore	Jhikargachha	Rajbari	Rajbari Sadar
Chattogram	Hathazari	Jashore	Sharsha	Rajshahi	Bagha
Chattogram	Lohagara	Jhenaidah	Jhenaidah Sadar	Rajshahi	Bagmara
Chattogram	Patiya	Jhenaidah	Kaliganj	Rajshahi	Charghat
Chuadanga	Alamdanga	Jhenaidah	Maheshpur	Rajshahi	Mohanpur
Chuadanga	Chuadanga Sadar	Kishoreganj	Austagram	Rajshahi	Paba
Chuadanga	Jibannagar	Kishoreganj	Bajitpur	Rajshahi	Puthia
Cox's Bazar	Chakaria	Kishoreganj	Bhairab	Rangpur	Pirgachha
Cox's Bazar	Ramu	Kishoreganj	Karimganj	Satkhira	Kalaroa
Cumilla	Adarsha Sadar	Kishoreganj	Kuliarchar	Sherpur	Nalitabari
Cumilla	Barura	Kishoreganj	Tarail	Sirajganj	Belkuchi
Cumilla	Brahmanpara	Kurigram	Chilmari	Sirajganj	Chouhali
Cumilla	Burichang	Kurigram	Kurigram Sadar	Sirajganj	Kazipur
Cumilla	Chandina	Kurigram	Rajarhat	Sirajganj	Shahjadpur

continued on next page

Table 8 *continued*

District	*Upazila*	District	*Upazila*	District	*Upazila*
Cumilla	Chauddagram	Kurigram	Roumari	Sirajganj	Sirajganj Sadar
Cumilla	Daudkandi	Kurigram	Ulipur	Sirajganj	Ullapara
Cumilla	Debidwar	Kushtia	Bheramara	Sunamganj	Chhatak
Cumilla	Laksam	Kushtia	Mirpur	Sunamganj	Derai
Cumilla	Lalmai	Lakshmipur	Ramganj	Sunamganj	Dharmapasha
Cumilla	Meghna	Magura	Magura Sadar	Sunamganj	Jagannathpur
Cumilla	Nangalkot	Magura	Mohammadpur	Sunamganj	Shalla
Cumilla	Titas	Magura	Shalikha	Sunamganj	Shantiganj
Dhaka	Keraniganj	Manikganj	Ghior	Sunamganj	Sunamganj Sadar
Dinajpur	Birol	Manikganj	Manikganj Sadar	Sylhet	Balaganj
Faridpur	Alfadanga	Manikganj	Shibalay	Tangail	Basail
Faridpur	Faridpur Sadar	Meherpur	Gangni	Tangail	Bhuanpur
Faridpur	Madhukhali	Moulvibazar	Moulvibazar Sadar	Tangail	Delduar
Faridpur	Sadarpur	Munshiganj	Munshiganj Sadar	Tangail	Kalihati
Faridpur	Saltha	Naogaon	Manda	Tangail	Mirzapur
Feni	Chhagalnaiya	Narail	Kalia	Tangail	Nagarpur
Feni	Daganbhuiyan	Narail	Lohagara	Tangail	Tangail Sadar
Feni	Feni Sadar	Narail	Narail Sadar	Thakurgaon	Haripur
Feni	Fulgazi	Narayanganj	Araihazar		

Source: Asian Development Bank consultants.

5.2.3 *Locations with Potential Level 3: Upazilas Where Caution Must Be Observed for Groundwater SIP Rollout*

The MCA identified 55 *upazilas* where local conditions must be carefully considered when rolling out SIP systems. These *upazilas* had a total MCA score lower than 2.5 (Figure 12 and Table 9). It is important to note that inclusion or non-inclusion on this list does not mean a particular SIP can or cannot be installed in these locations. Each individual location must be taken on its merits according to prevailing conditions at that site.

Figure 12: *Upazilas* Where Higher Caution Must Be Observed for the Rollout of Groundwater Solar Irrigation Pumps

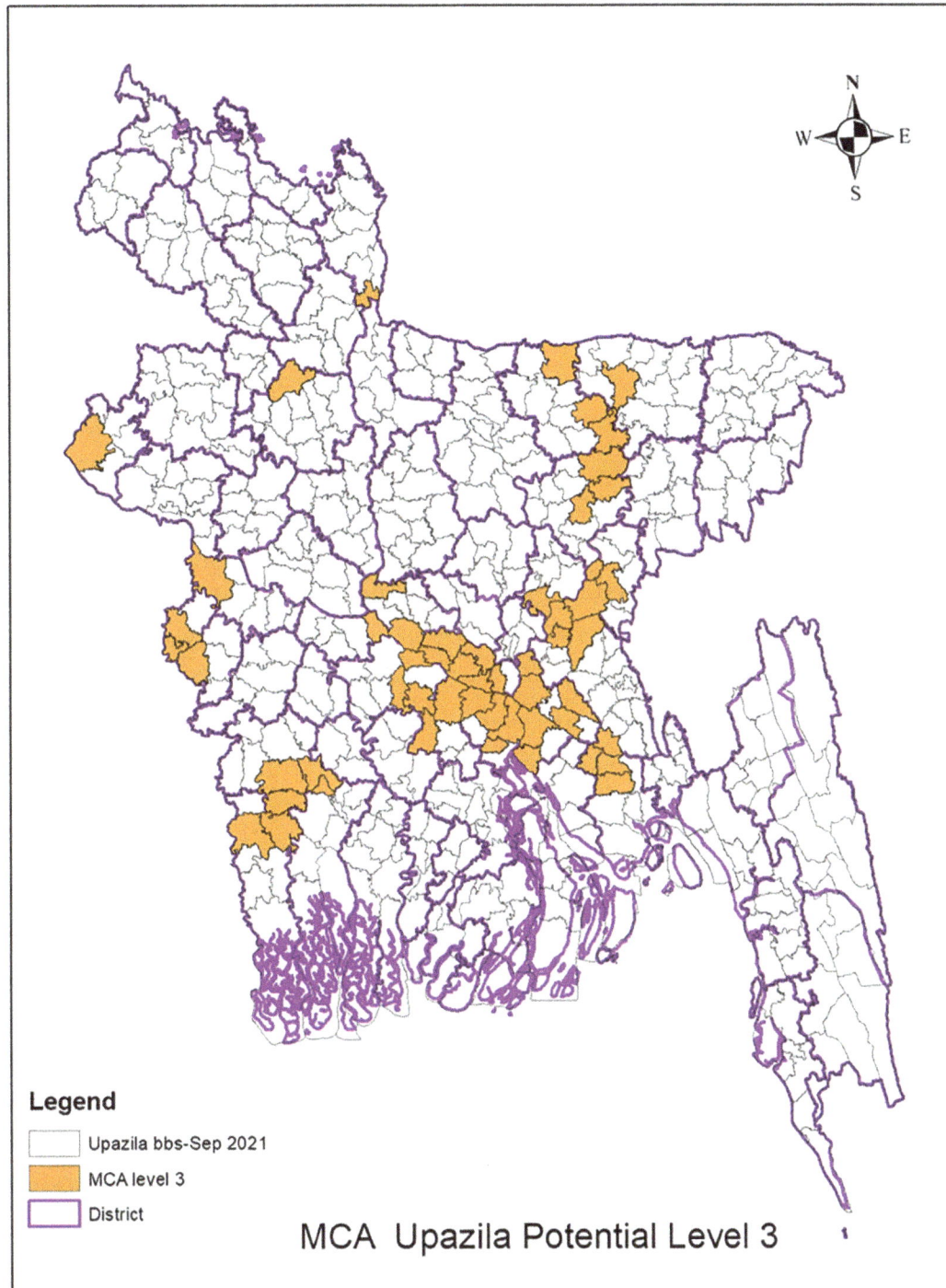

MCA Upazila Potential Level 3

Legend
- Upazila bbs-Sep 2021
- MCA level 3
- District

bbs = Bangladesh Bureau of Statistics, GWS = groundwater storage, MCA = multi-criteria analysis, TWS = terrestrial water storage.
Source: Asian Development Bank consultants.

Table 9: *Upazilas* Where Caution Must Be Observed for the Rollout of Solar Irrigation Pumps

District	*Upazila*	District	*Upazila*
Bogura	Shibganj	Kushtia	Daulatpur
Brahmanbaria	Ashuganj	Madaripur	Rajoir
Brahmanbaria	Banchharampur	Madaripur	Shibchar
Brahmanbaria	Brahmanbaria Sadar	Manikganj	Daulatpur
Brahmanbaria	Nabinagar	Manikganj	Harirampur
Chandpur	Chandpur Sadar	Meherpur	Meherpur Sadar
Chandpur	Haimchar	Meherpur	Mujibnagar
Chandpur	Hajiganj	Munshiganj	Gazaria
Chandpur	Kachua	Munshiganj	Louhajang
Chandpur	Matlab Uttar	Munshiganj	Sirajdikhan
Chapainababganj	Shibganj	Munshiganj	Sreenagar
Chuadanga	Damurhuda	Munshiganj	Tongibari
Cumilla	Homna	Narsingdi	Narsingdi Sadar
Cumilla	Manoharganj	Netrakona	Kalmakanda
Cumilla	Muradnagar	Netrakona	Khaliajuri
Dhaka	Dohar	Netrakona	Mohanganj
Dhaka	Nawabganj	Noakhali	Begumganj
Faridpur	Bhanga	Noakhali	Chatkhil
Faridpur	Char Bhadrasan	Noakhali	Sonaimuri
Faridpur	Nagarkanda	Rajbari	Goalanda
Jashore	Abhaynagar	Satkhira	Satkhira Sadar
Jashore	Keshabpur	Satkhira	Tala
Jashore	Manirampur	Shariatpur	Bhedarganj
Khulna	Phultala	Shariatpur	Damudya
Kishoreganj	Itna	Shariatpur	Naria
Kishoreganj	Mithamain	Shariatpur	Zajira
Kishoreganj	Nikli	Sunamganj	Jamalganj
Kurigram	Rajibpur		

Source: Asian Development Bank consultants.

5.3 *Upazilas* Where SIP Rollout for Groundwater Is Not Recommended

The MCA excluded 145 *upazilas* for which the Government of Bangladesh has not provided guidance for tube well distance. These are not recommended for SIP rollout. These *upazilas* are mostly along the coast, subject to tides or floods, with groundwater quality unsuitable for irrigation (Figure 13 and Table 10). These locations include, for example, saline-affected and flood-prone areas. The installation of SIP systems in hilly, flood-prone, or saline-affected areas is 30%–50% more expensive than regular installations.[57] For saline-affected areas, water purification should be considered. Therefore, it is much more expensive to replace diesel pumps in these areas, and the return on investments is also much slower. For flood-prone areas, PV panels, the inverter, and the pump must be installed above the highest flood level, which requires extra civil engineering works. It is important to note

[57]   ADB consultants estimate.

that the inclusion or non-inclusion in the selection of an *upazila* does not mean a particular SIP can or cannot be installed in these locations. Each individual location must be taken on its merits according to prevailing conditions at that site.

Figure 13: *Upazilas* Where Solar Irrigation Pump Rollout for Groundwater Is Not Recommended

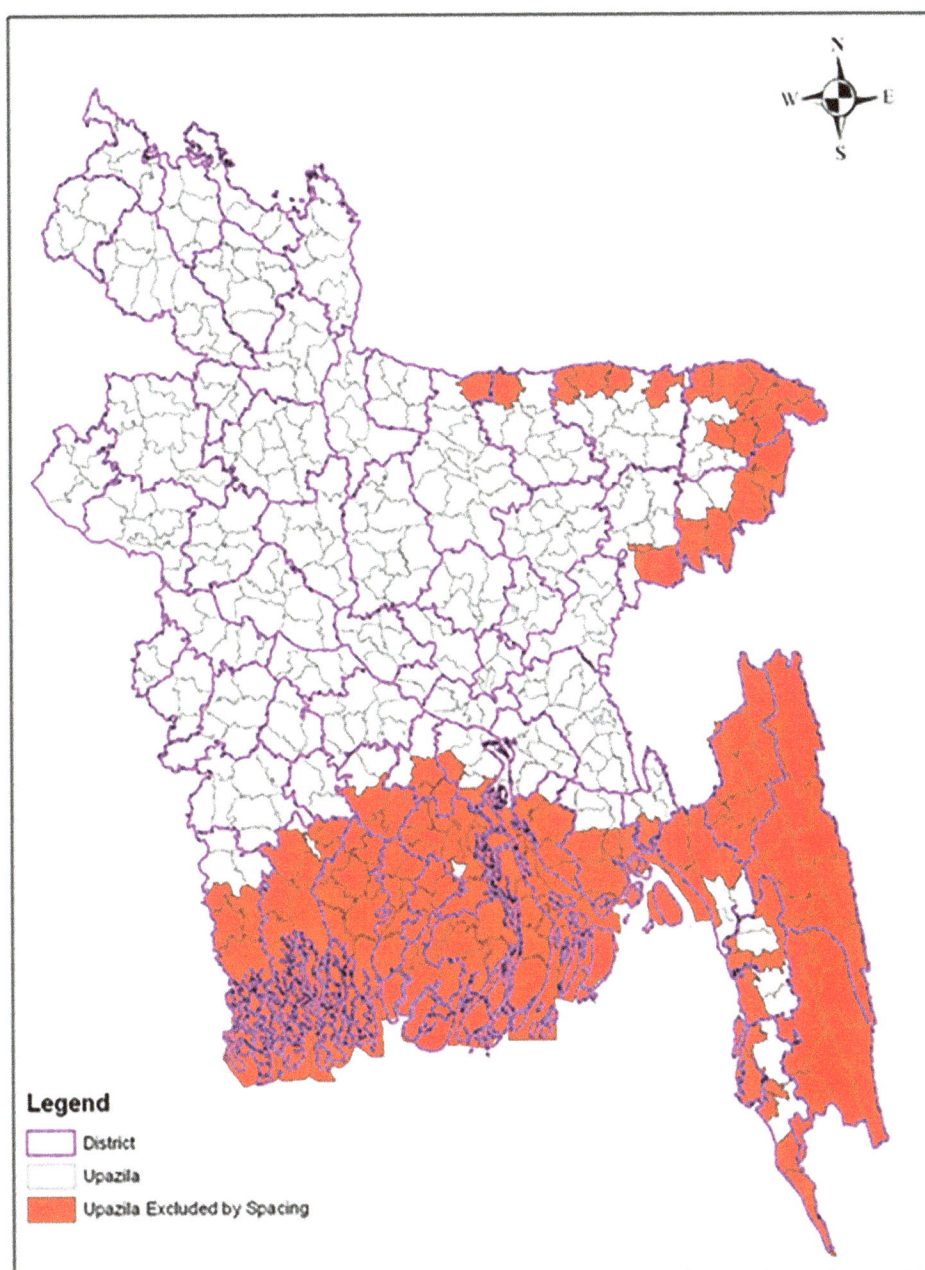

Legend

- District
- Upazila
- Upazila Excluded by Spacing

Source: Asian Development Bank consultants.

Table 10: *Upazilas* Not Recommended for Groundwater Solar Irrigation Pumps Rollout

Division	District	Upazila	District	Upazila	District	Upazila
Barishal	Barguna	Amtali	Barishal	Muladi	Patuakhali	Dashmina
	Barguna	Bamna	Barishal	Ujirpur	Patuakhali	Dumki
	Barguna	Barguna Sadar	Bhola	Bhola Sadar	Patuakhali	Galachipa
	Barguna	Betagi	Bhola	Borhanuddin	Patuakhali	Kalapara
	Barguna	Patharghata	Bhola	Charfasson	Patuakhali	Mirzaganj
	Barguna	Taltali	Bhola	Daulatkhan	Patuakhali	Patuakhali Sadar
	Barishal	Agailjhara	Bhola	Lalmohan	Patuakhali	Rangabali
	Barishal	Babuganj	Bhola	Monpura	Pirojpur	Bhandaria
	Barishal	Bakerganj	Bhola	Tazumuddin	Pirojpur	Indurkani
	Barishal	Banaripara	Jhalokati	Jhalokathi Sadar	Pirojpur	Kawkhali
	Barishal	Barishal Sadar (Kotwali)	Jhalokati	Kanthalia	Pirojpur	Mathbaria
	Barishal	Gaurnadi	Jhalokati	Nalchhity	Pirojpur	Nazirpur
	Barishal	Hijla	Jhalokati	Rajapur	Pirojpur	Nesarabad (Swarupkathi)
	Barishal	Mehendiganj	Patuakhali	Bauphal	Pirojpur	Pirojpur Sadar
Chattogram	Bandarban	Alikadam	Cox's Bazar	Kutubdia	Lakshmipur	Ramgati
	Bandarban	Bandarban Sadar	Cox's Bazar	Maheshkhali	Noakhali	Companiganj
	Bandarban	Lama	Cox's Bazar	Pekua	Noakhali	Hatiya
	Bandarban	Naikkhongchhari	Cox's Bazar	Teknaf	Noakhali	Kabirhat
	Bandarban	Rowangchhari	Cox's Bazar	Ukhia	Noakhali	Noakhali Sadar
	Bandarban	Ruma	Feni	Sonagazi	Noakhali	Subarnachar
	Bandarban	Thanchi	Khagrachhari	Dighinala	Rangamati	Baghaichhari
	Chattogram	Anwara	Khagrachhari	Guimara	Rangamati	Barkal
	Chattogram	Banshkhali	Khagrachhari	Khagrachhari Sadar	Rangamati	Belaichhari
	Chattogram	Chandanaish	Khagrachhari	Lakkhichhari	Rangamati	Jurachhari
	Chattogram	Fatikchhari	Khagrachhari	Mahalchhari	Rangamati	Kaptai
	Chattogram	Karnaphuli	Khagrachhari	Manikchhari	Rangamati	Kawkhali
	Chattogram	Mirsarai	Khagrachhari	Matiranga	Rangamati	Langadu
	Chattogram	Rangunia	Khagrachhari	Panchhari	Rangamati	Naniarchar
	Chattogram	Sandwip	Khagrachhari	Ramgarh	Rangamati	Rajasthali
	Chattogram	Sitakunda	Lakshmipur	Kamalnagar	Rangamati	Rangamati Sadar
	Cox's Bazar	Coxs Bazar Sadar	Lakshmipur	Lakshmipur Sadar		
	Cox's Bazar	Eidgaon	Lakshmipur	Raipur		

continued on next page

Table 10 *continued*

Division	District	Upazila	District	Upazila	District	Upazila
Dhaka	Gopalganj	Gopalganj Sadar	Madaripur	Dasar	Shariatpur	Gosairhat
	Gopalganj	Kotalipara	Madaripur	Kalkini	Shariatpur	Shariatpur Sadar
	Gopalganj	Tungipara	Madaripur	Madaripur Sadar		
Khulna	Bagerhat	Fakirhat	Khulna	Batiaghata	Khulna	Terokhada
	Bagerhat	Kachua	Khulna	Dacope	Satkhira	Ashashuni
	Bagerhat	Mollahat	Khulna	Dighalia	Satkhira	Debhata
	Bagerhat	Mongla	Khulna	Dumuria	Satkhira	Kaliganj
	Bagerhat	Morelganj	Khulna	Koyra	Satkhira	Shyamnagar
	Bagerhat	Rampal	Khulna	Paikgachha	Bagerhat	Bagerhat Sadar
	Bagerhat	Sharankhola	Khulna	Rupsa	Bagerhat	Chitalmari
Mymensingh	Mymensingh	Dhobaura	Netrakona	Durgapur		
Sylhet	Habiganj	Chunarughat	Sunamganj	Bishwambharpur	Sylhet	Golapganj
	Habiganj	Shayestaganj	Sunamganj	Dowarabazar	Sylhet	Gowainghat
	Moulvibazar	Baralekha	Sunamganj	Madhyanagar	Sylhet	Jaintapur
	Moulvibazar	Juri	Sunamganj	Tahirpur	Sylhet	Kanaighat
	Moulvibazar	Kamalganj	Sylhet	Beanibazar	Sylhet	Sylhet Sadar
	Moulvibazar	Kulaura	Sylhet	Companiganj	Sylhet	Zakiganj
	Moulvibazar	Sreemangal	Sylhet	Dakkhin Surma		

Source: Asian Development Bank consultants.

5.3.1   Salinity-Affected Areas

Because the soil in coastal regions is salinized, farming practices and yields from agriculture vary. Bangladesh has seen significant sea level rises, with measurements showing rises of 4 mm annually at Hiron Point in the west, 6 mm annually at Char Changa in the country's center, and even 8 mm annually at Cox's Bazar in the southeast (footnote 29). One cause for the increase from 1.5 million hectares under mild salinity in 1973 to 3 million hectares under salinity in 2007 was these sea level rises.[58] Areas regularly affected by soil salinity are shown in Figure 14.

In addition, shrimp farming has come to predominate in coastal regions, which adversely affects potential farming patterns as well as cropping patterns. Since the salinity intensity increases during the Kharif I and Rabi seasons, the majority of farmers grow vegetables on their land for their own consumption.[59] Crops such as mango, jackfruit, blackberry, betel nut, and date palm are disappearing gradually. Farmers cultivate both local and high-yield variety Aman rice in the Kharif II season.

---

[58]   F. Khatun and A.K.M. Nazrul Islam. 2010. Policy Agenda for Addressing Climate Change in Bangladesh: Copenhagen and Beyond. *Occasional Paper. No. 88*. Dhaka: Centre for Policy Dialogue. http://www.cpd.org.bd/pub_attach/op88.pdf.

[59]   The Kharif season starts at the end of March and runs till November when the moisture from rainfall plus soil storage is enough to support rainfed crops. It is characterized by high temperatures, rainfall, and humidity. The season is divided into Kharif I (last week of March until May) and Kharif II (June–November). The Rabi season begins at the end of the humid period, when the southeast monsoon starts ending in October, and runs until end-March. Rabi is characterized by dry sunny weather, and is cool during December–February. The average length of the Rabi season is 100–120 days in the extreme west to 140–150 days in the northeast.

Figure 14: Groundwater Salinity Concentration in Bangladesh

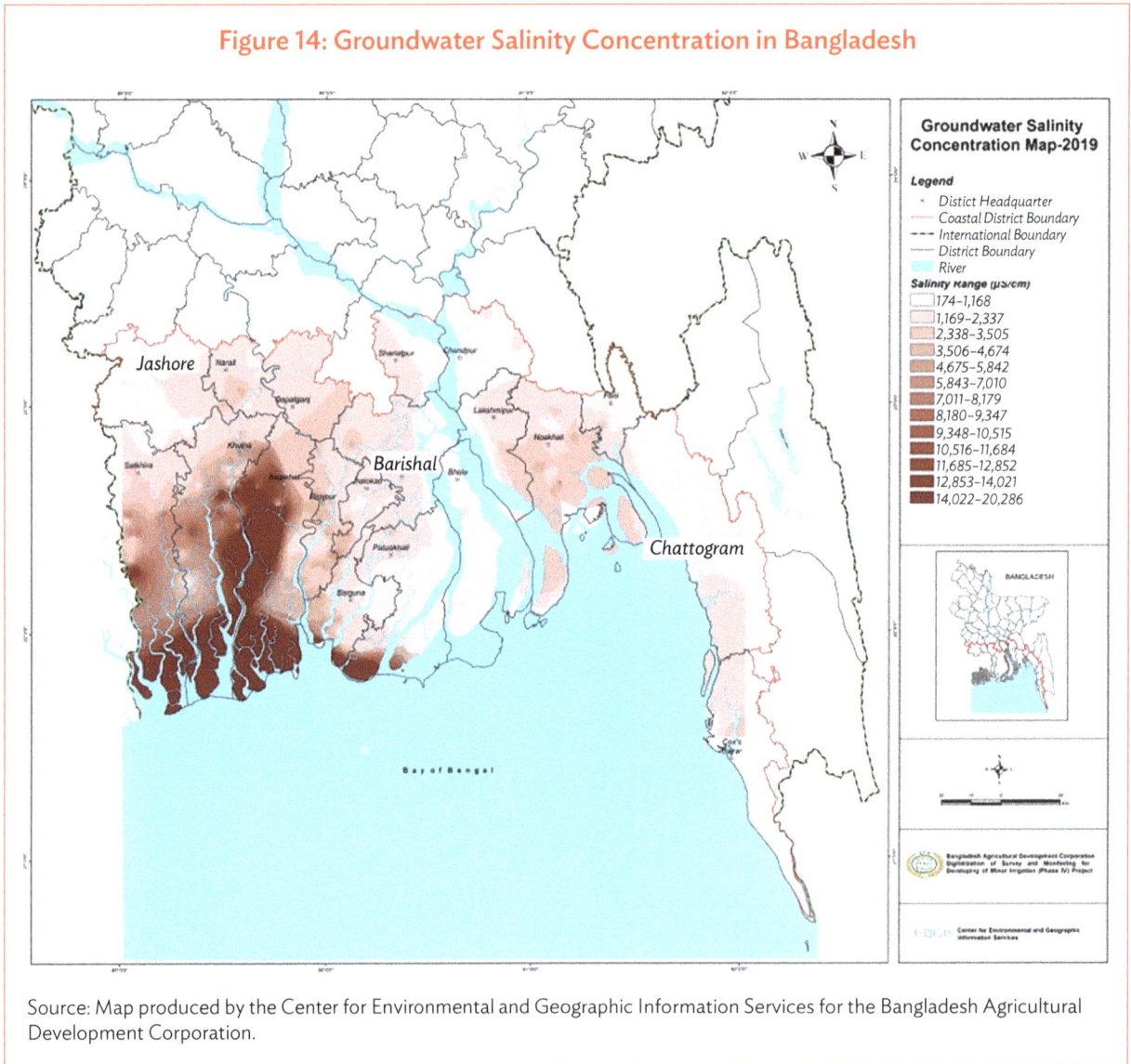

Source: Map produced by the Center for Environmental and Geographic Information Services for the Bangladesh Agricultural Development Corporation.

Land fertility is also impacted by shrimp farming. Salt percolates through the soil via gravity when saline water is stored for a longer period of time, which is necessary for shrimp cultivation. To make certain that shrimp grow more effectively during the monsoon, shrimp farmers also continuously add more salt to the water. The excess salt remains in the field, raising the salinity even further. Consequently, raising shrimp raises the salinity of the surrounding nonsaline soil, which has a negative impact on crop cultivation in the non-shrimping season. Because there aren't enough salt-tolerant varieties, vegetable and other crop yields decrease in shrimp farming areas.

5.3.2   Flood-Prone Areas

By the 2050s, Bangladesh is predicted to be 4% wetter (footnote 28). Compared to the 1970–2000 average, the mean annual rainfall is expected to rise by 7% on average by the 2090s, though some models predict increases of up to 24%. In terms of region, the north and northwest are predicted to see a larger increase than the south. Any particular year consists of monsoon and non-monsoon seasons. By the 2090s, the monsoon season will see the largest increases, with an average of 14%.[60] For the non-monsoon season, projections are mixed, with some models projecting rainfall decreases (footnote 60) and others projecting increases of about 10% by the 2090s.

Large increases are expected in 5-day rainfall maxima, especially during the wet season, and an increasing portion of total rainfall will fall during "heavy rainfall events," indicating a rainfall pattern with more extremes. The sea level will also rise further. According to the Intergovernmental Panel on Climate Change, there will be an increase (compared to 2000) of 14 cm by 2030, 32 cm by 2050, and 88 cm by 2100. Bangladesh has experienced a higher relative rise in sea level than many other nations because low-lying coastal areas have simultaneously been submerged. A significant number of farmers and their irrigation operations will be impacted by this. Around 27 million people will be at risk due to sea level rise by 2050.

Areas prone to severe flooding are in the Haor Basin of the northern belt of Bangladesh, which comprises Sylhet, Sunamganj, Moulvibazar, Habiganj, and Netrakona districts, as well as the southeast in Chattogram, Cox's Bazar, and Bandarban districts. Districts that have rivers running through them or are near the sea and are also prone to severe flooding are Brahmanbaria, Barishal, Bogura, Chandpur, Chapai Nawabganj, Cumilla, Dhaka, Faridpur, Gaibandha, Gazipur, Jamalpur, Kishoreganj, Kurigram, Kushtia, Lakshmipur, Madaripur, Manikganj, Munshiganj, Narayanganj, Narsingdi, Natore, Pabna, Rajbari, Rajshahi, Rangpur, Shariatpur, Sherpur, Sirajganj, and Tangail. The effects of river flooding in northern Bangladesh are often most severe due to preexisting vulnerabilities. Figure 15 provides an illustration of the severity of flooding in Bangladesh.

60   A. Karmalkara et al. 2012. *UNDP Climate Change Country Profiles: Bangladesh.* https://www.geog.ox.ac.uk/research/climate/projects/undp-cp/ UNDP_reports/Bangladesh/Bangladesh.hires.report.pdf.

Figure 15: Sample of Flooding Inundation Map of Bangladesh

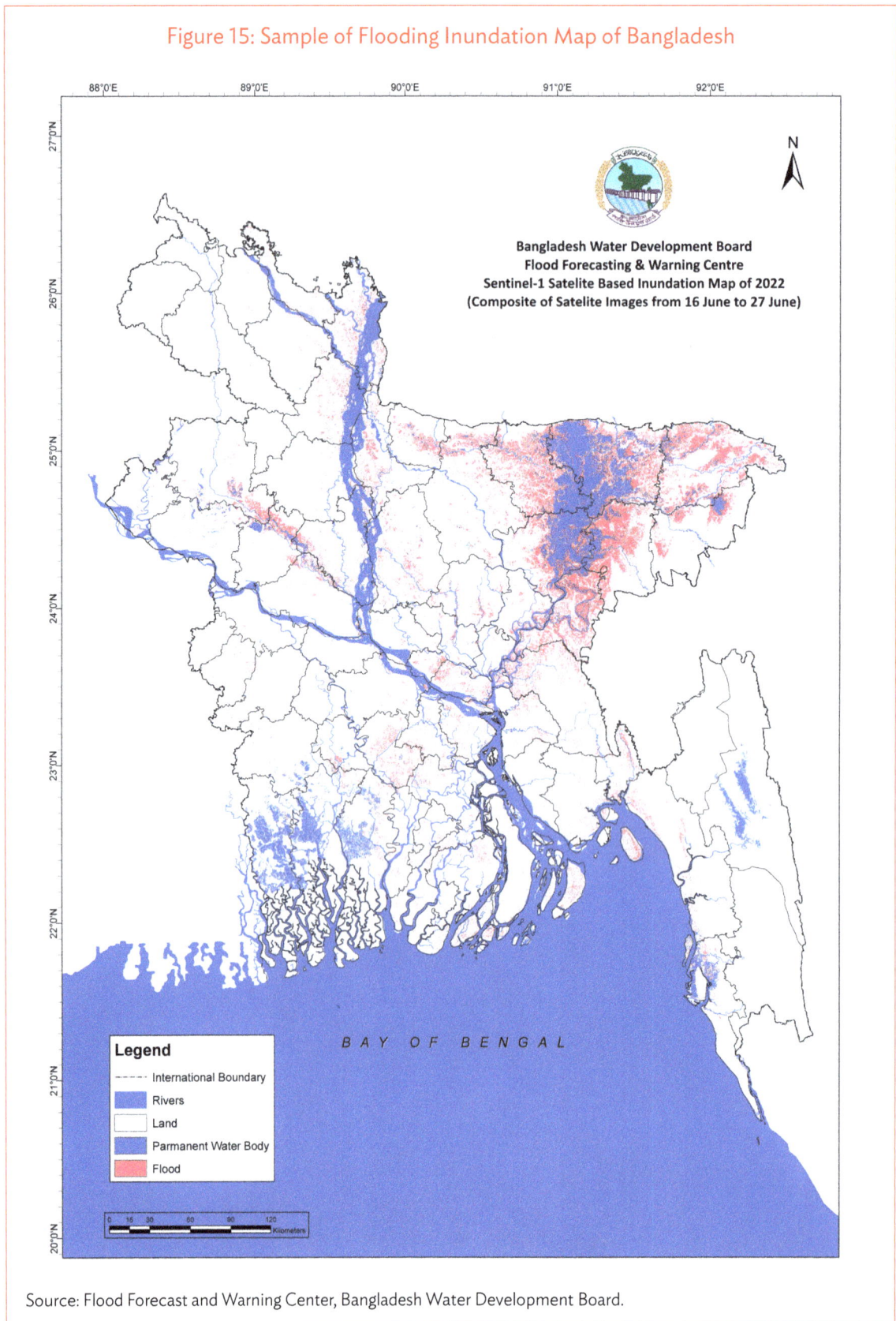

Source: Flood Forecast and Warning Center, Bangladesh Water Development Board.

6. State-of-the-Art Solar Irrigation in Bangladesh

6.1 Responsible Ministries

Bangladesh has more than 35 central government organizations, affiliated with about 13 ministries or divisions, that are working on water sector planning, development, and management involving water management projects related to irrigation, flood management, and drainage (footnote 50). Inter-agency coordination and partnership are limited during planning, implementation, and post-implementation (e.g., monitoring and updating) phases for most water-related projects, which are multisector and multidisciplinary, and require an integrated approach from the relevant organizations to be most effective.

Cross-sectoral cooperation between the Ministry of Agriculture, Ministry of Water Resources, Ministry of Local Government, and MOPEMR is essential for a successful SIP rollout. Within the MOPEMR, the Power Division is responsible for the implementation of renewable energy in the country and for electricity management activities.

One way to enhance cross-sectoral cooperation is to have a high-level working group or task force set up with senior representatives from these agencies who have the power to make the decisions that will enable an effective and smooth SIP rollout. SREDA should be given a project coordination role to align efforts of all the different implementing agencies (BADC, BARI, BMDA, BWDA, DAE, six electricity distribution utilities,[61] IDCOL, LGED, RDA, and others using public funding).

6.2 Relevant Agencies

The relevant agencies based on their involvement, experience, and interest in SIP for irrigation are listed in Table 11. The private sector also installs tube wells for irrigation and other purposes but is not involved in irrigation management.

---

[61] Bangladesh Power Development Board (BPDB), Bangladesh Rural Electrification Board (BREB), Dhaka Electricity Supply Company (DESCO), Dhaka Power Distribution Company (DPDC), Northern Electricity Supply Company (NESCO), and West Zone Power Distribution Company Limited (WZPDCL). These electricity distribution companies will be in charge of ensuring grid integration of future solar irrigation pump systems.

Table 11: Relevant Agencies for Irrigation with Solar Irrigation Pump Systems

Agency	Ministry	Expertise
BADC	Agriculture	SIP systems in DTWs, STWs, and LLPs with standalone solar pumps and with net metering. BADC operates 22% of DTW pumps in Bangladesh.
BMDA	Agriculture	SIP systems in DTWs, STWs, and LLPs with standalone solar pumps and with net metering. BMDA operates 42% of DTW pumps in Bangladesh.
BREB and Other Distribution Utilities	Power Division	New installations of tube wells, coupled with submersible standalone solar pumps with export facility and solar-electric with net metering (import and export).
BWDB	Water Resources	Responsible for surface water and groundwater management, and has relevant experience in supporting solar irrigation projects. BWDB must be consulted before selecting sources for surface water irrigation and aquifers for groundwater irrigation.
LGED	Local Government and Rural Development	Experienced in working with local-level institutions like Union Parishad and Zila Parishad. LGED could try to involve these local-level offices for the maintenance, operation, and management of LLP SIP projects at the rural level, especially in the south where the maximum numbers of LLPs operate during Boro season.
IDCOL	Ministry of Finance	New installations of tube wells, coupled with submersible standalone solar pumps with export facility and solar-electric with net metering (import and export).

BADC = Bangladesh Agriculture Development Corporation, BMDA = Barind Multipurpose Development Authority, BREB = Bangladesh Rural Electrification Board, BWDB = Bangladesh Water Development Board, DTW = deep tube well, LGED = Local Government Engineering Department, IDCOL = Infrastructure Development Company Limited, LLP = low-lift pump, SIP = solar irrigation pump, and STW = shallow tube well.

Source: Asian Development Bank consultants.

6.3   Pilot Interventions

Previous interventions represent a learning period that can be leveraged for a successful SIP rollout. A number of programs implemented since 2009 (Table 12) plan to complete the installation of at least an additional 3,800 SIP systems in the short term. All these interventions aim to develop institutional capacity and demonstrate the economic benefits of SIP systems to farmers.

Table 12: Characteristics of Some Pilot Programs for Solar Irrigation Pump Systems Recorded, 2009–September 2022

Program	Installed		Pump Ratings	Financing-Working Model and Target Users (Where Information Available)
	Units	kWp		
BADC	272	1,989	Most are 1.5, 2.2, 4, 7.5, and 11 kW. Two pumps are 35 kW	Irrigation and drinking water. Additional 1,000 units planned by 2023. Aimed at groups of small and very small farms. Upfront 65% grant, rest financed by BADC.
BARI	37	510	Average 7 kW	BARI's solar irrigation program.
BMDA	119	2,427	11 kW	Irrigation. Fee-for-service model (includes management, operation, and maintenance) with 100% grant. Aimed at serving small and very small farms.

continued on next page

Table 12 *continued*

Program	Installed		Pump Ratings	Financing-Working Model and Target Users (Where Information Available)
	Units	kWp		
BMDA	390	1,560	2.2–3.7 kW	Irrigation and drinking water. Fee-for-service model (includes management and O&M) with 100% grant. Aimed at serving small and very small farms.
BREB	40	200	3.7 kW	Irrigation. 100% financed in 2010–2011, with 20 pumps financed by KOICA and 20 pumps by the Clean Climate Trust Fund through BREB. Installation managed by users.
BREB (financed by ADB)	150	1,298	3.0–15 kW	Irrigation. 2,000 SIP systems planned (ongoing implementation) in 20 districts, financed by ADB. Ownership model, with 50% grant, O&M services included in project cost. Addressed to small and very small farms.
DAE	40	304	4 kW	Irrigation. Ownership model with groupings of 25 farmers. Fully financed by the government. Projects include flooded and hilly areas.
GIZ	122	N/A	N/A	Drinking water. Financed by GIZ with 100% grant. Management by users. Provided to population with no clean drinking water supply.
IDCOL	1,531	40,330	3.7–22 kW, most 18.5 kW	Irrigation. Target of installing up to 10,000 SIP systems by 2030. Fee-for-service model with 50% grant. Aimed at small and medium-sized farms, likely with some agribusiness activities. O&M service not included in project cost. Ownership model will also be offered for upcoming projects.
RDA	17	204	Most 7.5 kW	Irrigation and drinking water. Ownership model. O&M (provided by RDA) is included in the share of project cost transferred to farmers. Farmers pay upfront 10% of this cost, rest paid in 15 years.
Total	2,718	48,822		

ADB = Asian Development Bank, BADC = Bangladesh Agriculture Development Corporation, BARI = Bangladesh Agriculture Research Institute, BMDA=Barind Multipurpose Development Authority, BREB = Bangladesh Rural Electrification Board, DAE = Department of Agriculture Extension, GIZ = Deutsche Gesellschaft für Internationale Zusammenarbeit, IDCOL = Infrastructure Development Company Limited, KOICA = Korea International Cooperation Agency, kW = kilowatt, kWp = kilowatt peak, N/A = no available data, O&M = operation and maintenance, RDA = Rural Development Academy, SIP = solar irrigation pump.

Source: Information collected in September 2022 from BADC, BARI, BMDA BARI, BREB, DAE, GIZ, IDCOL, and RDA.

IDCOL offers concessionary credit, capacity development support, and startup subsidies for projects and programs that make use of renewable energy sources. IDCOL's objective is to ensure financial and economic sustainability of their projects with a goal of commercialization. IDCOL will culminate its flagship program of 10,000 SIP systems by 2030.

The BREB has initiated a flagship project that aims to install 2,000 SIP systems to provide strong demonstration effects across the country. This pilot program is financed with a $20 million ADB loan, a $22.44 million grant from the Scaling-Up of Renewable Energy Program of the Strategic Climate Fund, and a $3 million output-based aid grant from the Clean Energy Financing Partnership Facility.

The DAE also aims to complete a flagship project installing an additional 500 SIP systems, though its financing and schedule have not been decided yet.

6.4 Lessons Learned

Several pilot interventions to investigate the feasible potential of solar irrigation have taken place in Bangladesh in recent years. The first SIP system in Bangladesh was introduced in 2009 by the BADC in Vakurta Savar. The year after, BREB introduced a pilot grant-based program with support from the Bangladesh Climate Change Trust Fund. IDCOL also started its first trial with a grant in Shapahar, Naogaon District, in 2010. In 2012, BREB initiated 20 SIP projects with support from the Korea International Cooperation Agency and 20 more projects with the support from the Bangladesh Climate Change Trust Fund; BADC and IDCOL implemented new pilot programs in the same year. BADC started with the installation of 7 SIP projects with buried pipe networks funded by the government while IDCOL launched its first 100 SIP projects in 2012. BMDA, DAE, RDA, and Deutsche Gesellschaft für Internationale Zusammenarbeit have also all implemented their own pilots or small programs for agricultural irrigation and drinking water. As of August 2020, more than 2,500 SIP systems were installed in Bangladesh with an installed capacity exceeding 47 MWp. More than 1,500 of these systems, with a capacity of 40 MWp, have been installed by IDCOL under a fee-for-service financing-working model.[62]

These pilot interventions to date have highlighted challenges and identified possible ways to scale up SIP systems. Programs for the installation of solar home systems (SHSs) are also a valuable source of information. Some 5.8 million SHSs are installed in Bangladesh; IDCOL installed 4.4 million, while the remaining were financed by Test Relief, Kabikha (Food for Work) and the Ministry of Chattogram Hill Tracts Affairs.

Lessons learned through these interventions are described in the following section. As part of the "Off-Grid Solar— Investment Mobilization Implementation Road Map" study funded by the International Climate Initiative of the German Federal Ministry for the Environment, Nature Conservation, Nuclear Safety, and Consumer Protection, some of these findings were discussed during stakeholder consultations held between July 2018 and January 2019.

6.4.1 *Inadequate Identification of Targeted Beneficiaries*

The most important factor to consider when selecting a financing-working model is the size and number of farms that will be served by an SIP system. The financial capabilities of farmers will determine the level of financial assistance required and the level of private sector participation.

Crop cultivation in Bangladesh takes place in millions of tiny-to-small and some medium-sized farms, most operating just 1–15 bighas of land (and some even less).[63] Such small farms account for almost 90% of all farms in the country. The differences among these small farms should have a strong impact on the final decision in selecting a financing-working model for the scale-up of SIP systems.

Very Small Farms

Very small farms usually operate less than 2 bighas of land and focus on subsistence farming. Very small farms and their farmers are at a greater risk of poverty than larger farms, and they usually have very limited financial resources. This severely hampers their efforts to improve productivity and increase income. BADC targets these

---

[62] Under a fee-for-service model, a sponsor owns the SIP system and provides irrigation services to farmers against a fee.
[63] 1 bigha is 1337.8 square meters.

types of farms in their solar irrigation programs. The typical size of pumps for these projects is 5.5 kW and 7.5 kW, and they provide irrigation and drinking water to around 50 farmers per project.

If farms are too small, resulting in a farming group that requires an SIP system with less than 5 kW pump capacity, the project may not be economically viable. The costs per unit of irrigated area in such small projects become rather high and less competitive against the renting of diesel pumps. In such cases, it is more economic and efficient to design SIP systems that serve several farmers to make the required SIP system size larger than 5 kW pump capacity.

Small to Medium-Sized Farms

Small to medium-sized farms are those mostly operating more than 2 bighas of land. These farms usually complement subsistence farming with small agribusiness activities. A grouping of this type of farm will require SIP systems of a larger size, usually with pumps between 11 kW and 25 kW capacity.

Small and medium-sized farms usually have more financial resources compared to very small farms, and therefore have more ways to improve their productivity and increase their income. IDCOL targets these types of farms with a fee-for-service financing-working model, in which farmers pay for irrigation water service to a project sponsor that serves several farms. While this may be more expensive than owning and managing a project themselves through their irrigation committee, it frees up valuable time that can be used for other productive activities, such as fertilizing, weeding, and other maintenance, raising livestock, training for improved productivity, or simply for leisure.

Farmers' Groups and Irrigation Licensing

BWDB has proven that water management organizations (WMOs) are effective. BWDB has formed and institutionalized WMOs in more than 3,300 irrigation projects for participatory water management. Creating and using WMOs is recommended to help the road map rollout. Farmers' groups usually have 15–50 farmers, with this number depending on the area of the extension of land each individual farmer has, the area of land to be irrigated, and the amount of water required. Farmers' groups, cooperatives, or associations require licenses issued by their *Upazila* Irrigation Committee to operate their tube wells for irrigation. An *Upazila* Irrigation Committee comprises 14–16 appointed officers, including farmers' representatives (Table 13).

Table 13: Composition of *Upazila* Irrigation Committees

No.	Composition of Members	Position
1	Chairman of *Upazila* Parishad (Nirbahi Officer)	Chairperson
2	*Upazila* Agricultural Officer	Member
3	Engineer from LGED	Member
4	*Upazila* Rural Development Officer	Member
5	Representative of BWDB	Member
6	Representative of BRDB or, in the field, a representative of BREB	Member
7	Representative of DPHE	Member
8	Representative of the *Upazila* Farmers Organization, nominated by the *Upazila* Disposal Officer	Member
9	Assistant Engineer, BADC/BMDA	Member Secretary

BADC = Bangladesh Agriculture Development Corporation, BMDA = Barind Multipurpose Development Authority, BRDB = Bangladesh Rural Development Board, BREB = Bangladesh Rural Electrification Board, BWDB = Bangladesh Water Development Board, DPHE = Department of Public Health Engineering, LGED = Local Government Engineering Department.
Source: Government of Bangladesh, Ministry of Agriculture. 2019. *Groundwater Management Rules for Agricultural Works*. Dhaka.

The *Upazila* chairman may allow other concerned persons to be members of the committee. The committee is convened by the member secretary at a place, date, and time designated by the chairman. The chairman, or in their absence the member designated by them, presides over committee meetings. At least five members are required for reaching quorum of the committee meeting.

The *Upazila* Irrigation Committee is responsible for processing, inspecting, verifying, and approving all irrigation license applications. The committee has the power to temporarily or permanently suspend any license that is in violation of its terms. The committee verifies the following issues in each irrigation license application:

- soil level status;
- distance of the closest tube well;
- the benefited area; and
- potential impacts on other tube wells.

A positive verification report is required to approve any tube well application. No licenses are given to irrigate areas smaller than 6 ha. Also, tube wells must respect a minimum distance requirement (Table 14).

Table 14: Example of Minimum Distance between Wells of Different Capacity (m)

Capacity	4.892 m³/day (2.0 cusec)	3.669 m³/day (1.5 cusec)	2.446 m³/day (1.0 cusec)	1.223 m³/day (0.5 cusec)
4.892 m³/day (2.0 cusec)	573	535	489	386
3.669 m³/day (1.5 cusec)	535	469	451	347
2.446 m³/day (1.0 cusec)	489	451	405	302
1.223 m³/day (0.5 cusec)	386	347	302	243

cusec = cubic feet per second, m = meter, m³ = cubic meter.

Notes:

1. Values vary from district to district and are based on unutilized recharge in districts. Values correspond to the *upazila* of Shibganj in Bogura district. A discharge capacity of 0.5 cusecs in districts where groundwater development is high would have a higher spacing recommended. In districts where groundwater development is low, closer spacing is allowed. The spacing is a surrogate for unutilized recharge to groundwater.
2. 1 cusec = 0.028316847 cubic meters per second.

Source: Government of Bangladesh, Ministry of Agriculture. 2019. *Groundwater Management Rules for Agricultural Works*. Dhaka.

If there are disagreements among farmers and the committee, then the District Irrigation Committee will step in to resolve the issue (Table 15).

Table 15: Composition of District Irrigation Committees

No.	Composition of Members	Position
1	Deputy Commissioner of the District	Convener
2	Superintendent of Police in the District	Member
3	Deputy Director, District Agriculture Extension	Member
4	District Fisheries Officer	Member

continued on next page

Table 15 *continued*

No.	Composition of Members	Position
5	District Livestock Officer	Member
6	Executive Engineer, LGED	Member
7	Deputy Director, BRDB	Member
8	Executive Engineer, BWDB	Member
9	Executive Engineer, BPDB	Member
10	Executive Engineer, BREB or General Manager PBS	Member
11	Executive Engineer, BMDA (where applicable)	Member
12	Representative of DPHE	Member
13	Representative of MOEFCC	Member
14	District Cooperative Officer	Member
15	Farmers' Representative (nominated by District Commissioner)	Member
16	Assistant Engineer, BADC/BMDA	Member Secretary

BADC = Bangladesh Agriculture Development Corporation, BMDA = Barind Multipurpose Development Authority, BPDB = Bangladesh Power Development Board, BRDB = Bangladesh Rural Development Board, BREB = Bangladesh Rural Electrification Board, BWDB = Bangladesh Water Development Board, DPHE = Department of Public Health Engineering, LGED = Local Government Engineering Department, MOEFCC = Ministry of Environment, Forest, and Climate Change, PBS = Palli Bidyut Samities (subsidiaries of BREB).

Source: Government of Bangladesh, Ministry of Agriculture. 2019. *Groundwater Management Rules for Agricultural Works*. Dhaka.

A key limitation in implementing the SIP project has been the issue of tube well spacing. This spacing is given for all *upazilas* that are recommended for groundwater extraction and is based on a specified instantaneous discharge of water. To reduce delays in approving licenses for SIP systems, it is recommended to amend the 2019 Groundwater Management Rules for Agricultural Works to replace the specified instantaneous discharges to day discharges. This is more realistic and will facilitate calculating the number and capacities of SIP systems.

6.4.2   Poor Site Selection, System Size, and Licensing

Issues of project design that have caused the failure of some projects include overlapping areas with other pumps, poor site selection, poor workmanship, and poor or wrong selection of pump capacity. The result of these deficiencies was that some SIP systems did not meet the irrigation needs of the farms.

Constraints related to land use and proprietary rights may result in poor site selection. These constraints should be dealt with on an individual basis before a project is granted. Any project that is close to a possible grid connection point should be given priority, as it will be possible to sell excess electricity back to the national grid. Before grid connection for a pump is approved, it should be made mandatory to undertake verification of whether there is a valid license to operate the pump issued by the relevant *Upazila* Irrigation Committee.

6.4.3   High Investment Costs and Low Uptake by Farmers

Bangladesh produces 100 MWp of PV panels annually and exports them mostly to Nepal and the Philippines.[64] As of 2020, the government gives Bangladeshi PV manufacturers 10% of the value of their exported products as an incentive to strengthen the local PV panels industry. Tax exemptions are already provided by the government

---

[64] S. Islam. 2019. Bangladesh Extends Incentive Scheme for Domestic Solar Industry. *PV Magazine. 29 October*. https://www.pv-magazine.com/2019/10/29/bangladesh-extends-incentive-scheme-for-domestic-solar-industry/.

for the importation of solar raw materials. However, to further encourage farmers to adopt SIP systems, additional equipment cost reductions are necessary. These include the complete waiver of custom duties for all related equipment, the value-added tax on pumps, and the advanced income and trade taxes on solar panels and pumps. Equipment for PV systems is expensive and is a bottleneck to scale-up of SIPs. Potential measures to mitigate this challenge are to reexamine existing tax incentives to either eliminate or further reduce the value-added tax and import duties to bring down costs for components.

In some cases, the low uptake of a new project by farmers has been attributed to the benefits of this relatively new technology not being fully demonstrated to them. Implementing agencies, such as BADC, BMDA, BWDB, IDCOL, and distribution utilities, have a responsibility to demonstrate the benefits of SIP systems, including their O&M practices, and to ensure that adequate spare parts are available and accessible. Successful programs are based on having good quality products and services, strong awareness among beneficiaries, good O&M practices, and good marketing. BREB is implementing a project to install 2,000 SIPs in 32 districts financed by ADB and the Scaling-Up of Renewable Energy Program. For this project, BREB is already undertaking awareness campaigns targeted at farmers to demonstrate the benefits of SIP systems. These awareness campaigns need to be taken up by all agencies involved with promoting SIP systems.

6.4.4 High Operation Costs and Low Utilization after Boro Season

When manual labor is involved and theft prevention measures are required, operations can become costly. High operating costs have been noted by project sponsors as a significant investment barrier while they gain experience with the technology. Sponsors must estimate properly the time for construction and commissioning, ensuring that the pump is operational in time for the Boro growing season to maximize revenues and reduce project risks. If farmers are not shown the full value right away, they will be hesitant to convert from diesel-powered pumps to SIPs. Switching pumps comes with costs, even though the tariffs offered will be on par with or less expensive than the cost of diesel-powered irrigation. An SIP system project's initial season's irrigation tariffs could be lowered to increase customer interest and engagement.

In January through April, during Boro season, there is a peak demand for irrigation. SIP systems are underutilized for the remainder of the year because crops in other seasons require less irrigation, or in some cases the SIP systems are simply unused if farmers decide to have no more than one crop per year. Hence, less than 50% of the pumps' technical power potential is being utilized. This is different from other countries, where demand for irrigation is more stable throughout the year, so revenue streams from pump operation are more regular.

High operation costs and low utilization may cause a mismatch between project cash flows and debt service obligations for the first few years when revenue generation is also typically at its lowest. The life of an SIP system is typically 20 years, but loans are typically for 10 years, though they may have grace periods (IDCOL gives a 2-year grace period). Project sponsors need to provide project-specific repayment profiles with increased repayment flexibility to prevent an early financial failure of the project. Increasing project returns and reducing the risk of seasonal demand would both be accomplished by generating income from the unutilized power generation. An alternative to selling surplus power to the grid is to use the PV panels and land around them to generate revenue through other activities, such as growing other crops under the shade of the PV panels or using the surplus electricity generated by the PV panels for purposes such as milling food or refrigeration.

6.4.5   Insufficient Provisions for Operation and Maintenance

The experience of BREB in installing 20 SIP systems financed by Korea International Cooperation Agency showed that insufficient or inadequate provisions for O&M was an important factor in the project's failure. These are features that should be clearly addressed upfront in an SIP program design and explicitly included in the contractual agreements for each project. Comprehensive warranties should be granted to each project during a period and specific monetary provision should be made after the warranty period (usually 5 years). After-sale warranty agreements must also be specific in relation to the time frame set to provide technical services. To avoid such situations, implementing agencies should include strong technical support features for O&M in the design of their programs.

In addition, all projects must have a dedicated person in charge of providing basic maintenance of the systems before and after the warranty period. If a circuit breaker trips or a solar PV panel is broken, the farmer needs to call the supplier's electrician, who may be based in Dhaka. It may take the technician 2–3 days, or even longer, to arrive at the site to fix the problems. If this happens during the peak irrigation period, farmers may lose their crops or revert to using diesel pumps to irrigate their fields. Even simpler incidents, like a rat biting into an AC/DC wire at night, will require a trained person to solve the problem immediately. Otherwise, the next day the whole system does not work or only works partially. Having a trained person on site or within a few hours to address the problem will shorten service times.

6.4.6   Reduced Interest in Keeping Systems after Grid Connection

The experience with the installation of SHSs shows that as BREB advances with the government's mandate of electrifying rural households, many of those households that had previously installed SHSs stopped servicing their loans once they had access to grid power supply. Some household customers requested that the SHSs be removed on account of having grid power supply. Other recently grid-connected households continue to use their SHS as a back-up power supply because voltage drops on the distribution lines make the grid power supply unreliable. Since some households were not servicing the loans that they had received for the installation of their SHSs, they were also not getting the technical service needed from the suppliers. Several thousand SHSs were taken back by suppliers due to non-payment of loans.

A plan could be formulated by the government to aggregate these solar panels from SHSs and use them for SIP systems in rural areas or in providing clean drinking water in areas where poor farmers are located on a priority basis.

6.4.7   Risk of Theft and Vandalization of Equipment

Installations require an adequate security system to preserve them from theft and vandalism. PV panels, inverters, and cabling are always at risk of being stolen. A security system around installations must be designed according to the specific risks of the location. Developing strong ownership of the SIP systems by project recipients helps to maintain the effectiveness of the security system. Otherwise, beneficiaries may feel that the project belongs to the government, or to the agency installing it, and do not feel responsible for keeping the security system in place and effective all the time.

6.5 Environmental Assessment

Although SIP systems have been hailed as the greener option for pumps that run on the grid or fossil fuels, if this type of equipment is to be genuinely sustainable, care must be taken in its application.

6.5.1 Major Environmental Threats

Irrigation with groundwater using diesel pumps in Bangladesh faces two large environmental threats: depletion of groundwater and contamination of groundwater and soil.

Depletion of Groundwater

In intensively irrigated areas, groundwater levels can fluctuate 5–15 m below ground level and, in some places, 5–23 m during peak irrigation season (footnote 50). Declining rates are highest (exceeding –0.5 m per year) in and around Dhaka City and the High Barind Tract region and high (0 to –0.05 m per year) in areas south of the Ganges River. In coastal areas, shallow groundwater levels show stable to slightly rising trends (0 to +0.1 m per year).

An examination of variations in groundwater levels in Barind revealed that the degree of the fluctuations grew with an increase in the number of DTWs and the amount of water abstracted (footnote 52). The BRAC found a declining trend in the groundwater table in the northwest over 30 years (1981–2011) (footnote 53). The most severely depleted district has been identified as Rajshahi, followed by Pabna, Bogura, Dinajpur, and Rangpur. The decline in the groundwater table was between –2.3 and –11.5m during the study period.

In Bangladesh, the depth of tube wells might range from 30 to 300 m. In the north and northwest, the depth may vary from 30 to 150 m, whereas in coastal and southern areas, or in the center of the country, tube well depth can vary from 100 to 300 m due to salinity. This means that in many areas it is possible to use DTW configurations, which is also preferred to STW configurations, to avoid the rapid depletion of shallow water resources and the contamination of crops and drinking water with arsenic.[65]

Water and Soil Contamination

Water contamination. Arsenic in groundwater is a major health concern in Asia and is a major risk in using STWs. In Bangladesh, shallow aquifers are the main source of drinking water and irrigation, especially during the dry Boro season. Part of the shallow aquifer contains arsenic concentrations above the national drinking water standard of 0.050 mg per L, particularly in the south and southwest; thousands of people around the country have been diagnosed with arsenicosis. The most well-known concern is arsenic entering the food chain, affecting food safety. This poses a potential dietary risk to human health in addition to the risk from drinking contaminated groundwater. Continuous buildup of arsenic in the soil from arsenic-contaminated irrigation water may reduce crop yields and also degrade the quality of soils, thus also affecting the nutritional value of agricultural products and income of rural farming communities.[66]

---

[65] Guidelines are that ground tube wells should not go beyond 150 m depth. Groundwater below 150 m depth is arsenic-free and reserved for domestic use.

[66] Food and Agriculture Organization of the United Nations. 2006. *Arsenic Contamination of Irrigation Water, Soil and Crops in Bangladesh: Risk Implications for Sustainable Agriculture and Food Safety in Asia.* Bangkok.

Surface water and groundwater deeper than shallow water are in many cases not contaminated with arsenic, and their exploitation may put less pressure on available water resources. For these reasons, the government encourages the use of surface water as much as possible. However, surface water is scarce in 70% of the country during the dry season, and farmers have no choice but to use deeper groundwater.

It is recommended that all government-supported programs include the obligation to test for arsenic contamination and for shallow water depletion.

Soil contamination. Diesel and lubricant spills are the major sources of soil contamination when using diesel pumps. The replacement of diesel pumps with SIP systems addresses the problem of soil contamination.

6.5.2   Determining Availability and Safe Use of Groundwater

Determining the availability and safe use of groundwater is only possible through the construction of testing wells to measure its hydraulic characteristics. The depth of static water before pumping begins, the sustainable volume of water that can be pumped per minute or per hour, the variation in water depth at one or more constant pumping rates, and the recovery time of the water depth after pumping is stopped, are all measurements that should be taken in testing wells.

In a nutshell, the testing that is covered in this section measures the hydraulic properties of an average well. When testing a well, the pump and power unit should be able to run continuously for more than twenty-four hours at a steady and variable pumping rate. To ensure the accuracy of test results, it is recommended to repeat this test for a few days and longer tests for up to several weeks.

A well's safe yield can be ascertained by operating the pump at a rate that will result in no more than 50% of the maximum drawdown (Figure 16). Regardless of yield, a safe well capacity need to be set and maintained for that condition.

It is important that the equipment used for well testing is in good condition for an accurate test, since it is undesirable to have a forced shutdown during the test. The test pump should be large enough to test the expected capacity of the well, even though this may be far beyond the amount of water required and may exceed the capacity of the permanent well pump.

Figure 16: Tests for the Determination of Safe Yield of a Solar Irrigation Pump System

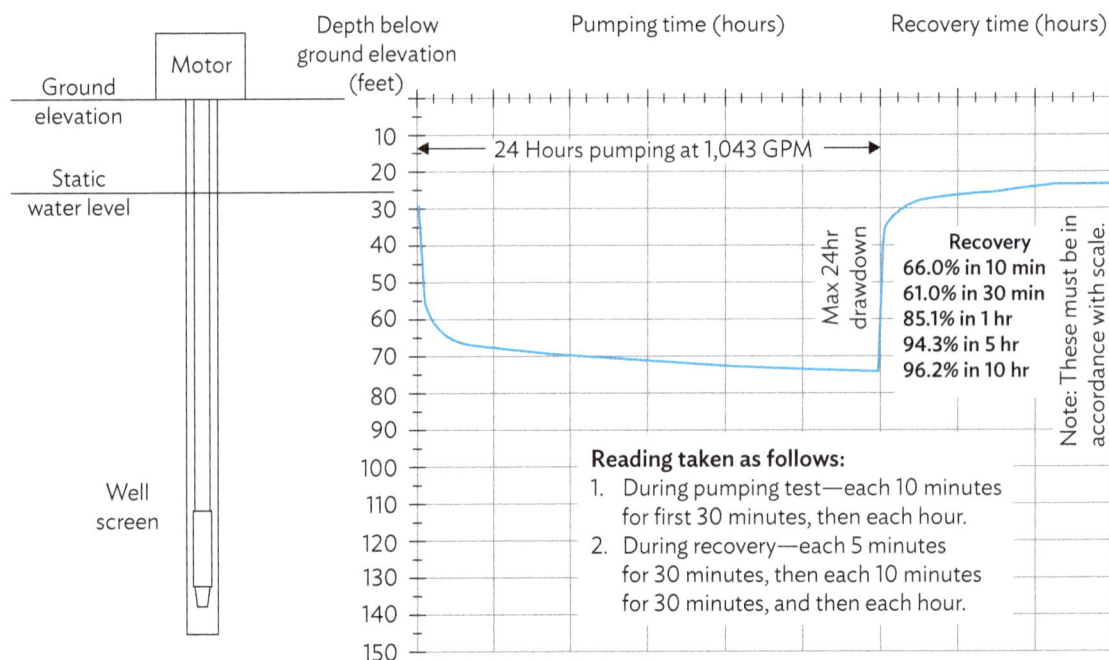

Source: Asian Development Bank consultants.

6.6 Geographic Preferences by Agency for Solar Irrigation Pump Rollout

The government's preference for prioritizing surface water for irrigation has been considered by each agency at their respective geographic location of sites for LLP and DTW SIP rollout at district level as shown in Table 16.

Table 16: Geographic Preferences for Low-Lift Pump and Deep Tube Well Solar Irrigation Pump Rollout per Agency at District Level

Agency	LLP	DTW	Mini Deep	Electricity Distribution Utility in Charge of Grid Integration
BADC	Sylhet	Sylhet		BPDB, BREB
	Sunamgonj	Sunamgonj		BPDB, BREB
	Moulvibazar	Moulvibazar		BPDB, BREB
	Habiganj	Habiganj		BPDB, BREB
	Netrokona	Netrokona		BPDB, BREB

continued on next page

Table 16 *continued*

Agency	LLP	DTW	Mini Deep	Electricity Distribution Utility in Charge of Grid Integration
	Kishoreganj	Kishoreganj		BPDB, BREB
	B'Baria	B.Baria		BPDB, BREB
	Cumilla	Cumilla		BPDB, BREB
	Narshingdi	Kushtia		BPDB, BREB
	Mymensingh	Chuadanga		BPDB, BREB
	Jamalpur	Jhenaidha		BPDB, BREB, WZPDCL
	Tangail	Jashore		BPDB, BREB, WZPDCL
	Sherpur	Magura		BPDB, BREB
		Narshingdi		BPDB, BREB
		Mymensingh		BPDB, BREB
		Jamalpur		BPDB, BREB
		Tangail		BPDB, BREB
		Sherpur		BPDB, BREB
BMDA	Thakurgaon	Thakurgaon		BREB, NESCO
	Panchagor	Panchagor		BREB, NESCO
	Dinajpur	Dinajpur		BREB, NESCO
	Nilphamari	Nilphamari		BREB, NESCO
	Kurigram	Kurigram		BREB, NESCO
	Lalmonirhat	Lalmonirhat		BREB, NESCO
	Rangpur	Rangpur		BREB, NESCO
	Gaibandha	Gaibandha		BREB, NESCO
	Joypurhat	Joypurhat		BREB, NESCO
	Bogura	Bogura		BREB, NESCO
	Sirajganj	Sirajganj		BREB, NESCO
	Pabna	Pabna		BREB, NESCO
	Natore	Natore	Natore	BREB, NESCO
	Rajshahi	Rajshahi	Rajshahi	BREB, NESCO
	C. Nawabganj	C. Nawabganj	C. Nawabganj	BREB, NESCO
	Naogaon	Naogaon	Naogaon	BREB, NESCO
BREB			Rangpur	BREB, NESCO
			Dinajpur	BREB, NESCO
			Thakurgaon	BREB, NESCO
			Panchagor	BREB, NESCO
			Bogura	BREB, NESCO

BADC = Bangladesh Agriculture Development Corporation, BMDA = Barind Multipurpose Development Authority, BPDB = Bangladesh Power Development Board, BREB = Bangladesh Rural Electrification Board, DTW = deep tube well, LLP = low-lift pump, NESCO = Northern Electricity Supply Company Limited, WZPDCL= West Zone Power Distribution Company Limited.

Source: Asian Development Bank consultants.

7. Financing-Working Models

Various financing-working models are needed to meet the needs of different clients who have a range of income and engage in varying types of agriculture activity. Solar irrigation in Bangladesh is constrained by a lack of proper irrigation facilities, production machinery, access to institutional credit facilities, difficulties in procuring inputs and storing products, and climate change. It must also compete with subsidized electricity. Consequently, the financing-working models used for the road map rollout must be flexible.

7.1   Classification and Main Characteristics of Financing-Working Models

Agencies such as the BADC, BMDA, BREB and other distribution utilities, BWDB, DAE, IDCOL, LGED, and RDA have sufficient interest and capability to support the rollout of SIP systems. Interviews conducted by the road map specialists with these agencies discussed the main characteristics of the financing-working models that could be implemented as part of the rollout (Table 17).

Table 17: Financing-Working Models Implemented by Leading Agencies

Agency	Managed by	Operated by	Maintained by	Repaired by	CAPEX Recovery	O&M Cost Recovery
BADC	Water users' cooperative	Water users' cooperative	Water users' cooperative and BADC	Minor by water users	No provision to realize	Follow different systems per hour and yearly basis
BMDA	Water users & BMDA	BMDA's appointed operator	BMDA	Medium & major repairs done by BMDA	Full realization	Full realization
BREB	Individuals or groups of buyers	Individuals or groups of buyers	Individuals or groups of buyers	Individuals or groups of buyers	Upfront payment, rest on installments	Individuals or groups of buyers
BWDB	Water users' groups	Community-based water users' group	Water users' groups	BWDB	No provision	Per unit irrigated area for different crops of different crop seasons
IDCOL	Private company or sponsor	Private company or sponsor	Private company or sponsor	Private company or sponsor	Upfront payment, rest on installments	Private company or sponsor

continued on next page

Table 17 *continued*

Agency	Managed by	Operated by	Maintained by	Repaired by	CAPEX Recovery	O&M Cost Recovery
LGED	Water users' groups	Community-based water users' groups	Water users' groups	Water users' groups	No provision	Experienced in working with local-level institutions like Union Parishad & other bodies

BADC = Bangladesh Agriculture Development Corporation, BMDA = Barind Multipurpose Development Authority, BREB = Bangladesh Rural Electrification Board, BWDB = Bangladesh Water Development Board, CAPEX = capital expenditure, IDCOL = Infrastructure Development Company Limited, LGED = Local Government Engineering Department, O&M = operation and maintenance.
Source: Asian Development Bank consultants.

These financing-working models can be categorized into the following, based on ownership and water source:

(i)   community/agency-based surface water irrigation (e.g., BADC/BMDA/WMOs);
(ii) community/agency-based groundwater irrigation (e.g., BADC/IDCOL/NGOs/private companies); and
(iii) individually owned groundwater irrigation (e.g., distribution utilities).

Each of these financing-working models has its merits and demerits.

Under community-based schemes, a group of farmers may reduce the number of tube wells and share project costs. Equity among the water users in community schemes is one of the primary considerations, across all different categories, such as rich and poor, landowners and lessees, and educated and uneducated. BMDA's prepaid metering system with a smart card for individual water users has to date successfully addressed cost recovery and equity in community-based schemes. The benefits of the prepaid metering system include the following:

• All water provided is paid for in advance, and there is no opportunity to bypass the meter.
• The system is completely transparent, with checks and balances to counter fraud.
• People cannot coerce the operator to deliver water free of charge.
• Poor farmers cannot be exploited by landowners who may control the well.
• The volume of water withdrawn is easy to estimate.

If an SIP is owned by individual farmers, as in the case of a pilot supported by ADB and executed by BREB, inequity in access to water does not arise.

Under individually owned schemes, the owner may also share irrigation services with their neighbors, a common practice in South Asia. In this case, they may also consider establishing a prepaid meter mechanism. Individual owners with grid-integrated SIPs may consider the following issues when they do a needs assessment:

• Coordination: What inter- and intra-departmental coordination mechanisms are needed?
• Affordability and financing: How can pumps be made affordable within state financial constraints?
• Targeting: How do the right-sized pumps reach the proper beneficiaries and locations?
• Infrastructure: What infrastructure can boost farmer benefits and reduce groundwater risks?
• Monitoring and evaluation: What aspects of schemes should be monitored, evaluated, and reported?

Adequate O&M are also critical factors for sustainable SIP management in all financing business models. A stronger post-rollout with solid technical support may be required by the individually owned schemes as learned from LGED and BREB pilot experiences.[67]

Regardless of the specific financing-working model for all of them, the government must provide grant financing to cover at least 50% of the benchmark cost of the SIP systems. The government may decide to increase its grant financing for specific target farmer groups, such as very small or poor farmers. Estimating the right value of the grant for those target farmer groups will require a specific national study to determine their payment capacity. The government should establish safeguards for farmers to repay loans in times of financial difficulties, such as in a year of poor harvest or after damage to crops caused by cyclones or other disasters. The government should also set aside funds for SIPs to be channeled through banks with low interest rates.

In addition to program administration to manage these financing-working models, implementing agencies must set up remote monitoring systems to monitor SIP performance after installation and implement consumer motivational and awareness activities to promote uptake of their programs. It is recommended that the government allocates a minimum of 2% of its grant financing annually for these agencies to defray their cost of carrying out remote monitoring, evaluations, and reports.

7.2 Modalities under Community-Based Groundwater Irrigation

Two framework modalities exist under this financing-working model:

(i) direct ownership by farmers and
(ii) fee-for-service by a private project sponsor, usually for a profit.

Both modalities have been applied in pilot interventions in Bangladesh.

7.2.1 Direct Ownership Modality

In the direct ownership modality, a group of farmers organized through an irrigation committee or cooperative owns and fully manages the operation of the SIP systems. Owners buy the equipment through a project sponsor (optional) who also helps them to process the subsidy and loan application with the financial institution (Figure 17).

This modality is based on the community-based natural resources management approach, which aims to create communities that are responsible for managing natural resources sustainably. In 2000, Bangladesh established its Guidelines for Participatory Water Management to ensure that all people could influence water decisions that affect them. The creation of elected community WMOs however placed restrictions on which local stakeholders have a say in water management. Limiting the representation of local stakeholders to WMOs has proved problematic, because often WMOs tend to lack both transparency and accountability.

This modality requires the government to increase the grant amount awarded to projects and improve the repayment conditions for small and very small subsistence farms.

---

[67] ADB. Promoting and Scaling Up Solar Photovoltaic Power through Knowledge Management and Pilot Testing in Bangladesh and Nepal. https://www.adb.org/projects/49103-001/main.

Figure 17: Structure and Organization of the Ownership Modality

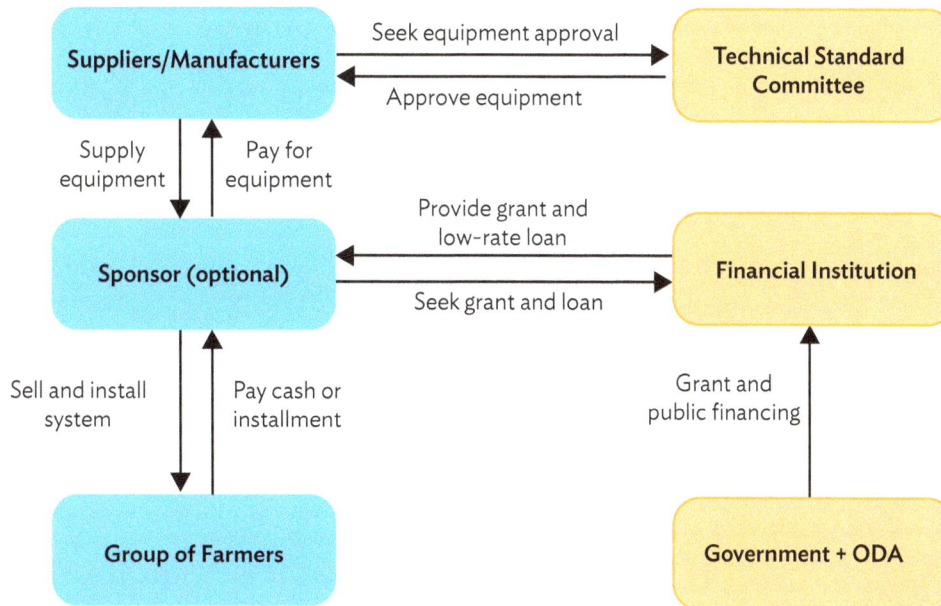

ODA = official development assistance.
Source: Modified and adapted from official presentations by the Infrastructure Development Company Limited (IDCOL).

Key Advantages of Direct Ownership Modality

(i)   **Extra motivation for multiple crops per year.** Since there is no variable operation cost when using an SIP system (no diesel consumption), small, poorer farmers are more motivated to plant a second or third crop during the year. Farming groups in many BADC projects charge only a single flat fee for the main irrigation season and will give the water for free to their members for the other two seasons.

(ii) **Fast acceptance of the technology.** Farmers quickly realize the many advantages of SIP systems, among them being time and money saved, since there is no need to travel to rent diesel pumps or buy diesel.

Bangladesh Agriculture Development Corporation (BADC) Solar Irrigation Pump Project. Project funded by BADC in Lalmonirhat under an ownership modality (photo provided by Asian Development Bank consultant).

(iii) **Drinking water and cold storage as additional use.** SIP systems can be used to produce clean drinking water or provide cold storage facilities to farmers, providing an additional benefit in places where sanitation infrastructure is poor or nonexistent, and improving the economic feasibility of the SIP project. Villages switching to clean water under this system have reported improvements in health, such as reduced cases of diarrhea and stomach ailments.

Specific Recommendations from Lessons Learned in Pilot Interventions

(i)   **Require a symbolic down payment for small farmers.** Projects designed to service very small farms will require a large part of the project cost to be subsidized since the target farmers do not have savings for any equity or relevant down payments, and their creditworthiness to get loans is very low. BADC currently claims a low (symbolic) down payment and gives a 20-year loan at 0% interest for the non-subsidized part.

(ii)   **Make sound arrangements for O&M.** Farmer groups are not always sufficiently trained in managing and maintaining SIP systems. It is strongly recommended to build training capacities across the country and to ensure that local technical support is available as part of the project design.

(iii)   **Ensure adequate agricultural extension services complement each project.** Farmers will need assistance in building up their knowledge in certain areas, such as in regard to the types of crops that can be grown, access to seed banks, improving their productivity, and commercializing their products. Cleaner and more efficient irrigation makes more sense when all other agricultural technical aspects are also well taken care of. Agricultural extension services should be intensified or coupled together with the implementation of an SIP system project.

7.2.2   Fee-for-Service Modality

In the fee-for-service modality, an intermediary (or project sponsor) owns and manages the SIP systems for a profit. The sponsor sells the irrigation water to farmers nearby and deals directly with equipment suppliers and the financial institutions that will provide the needed private loan and public subsidy (Figure 18).

Figure 18: Structure and Organization of the Fee-for-Service Modality

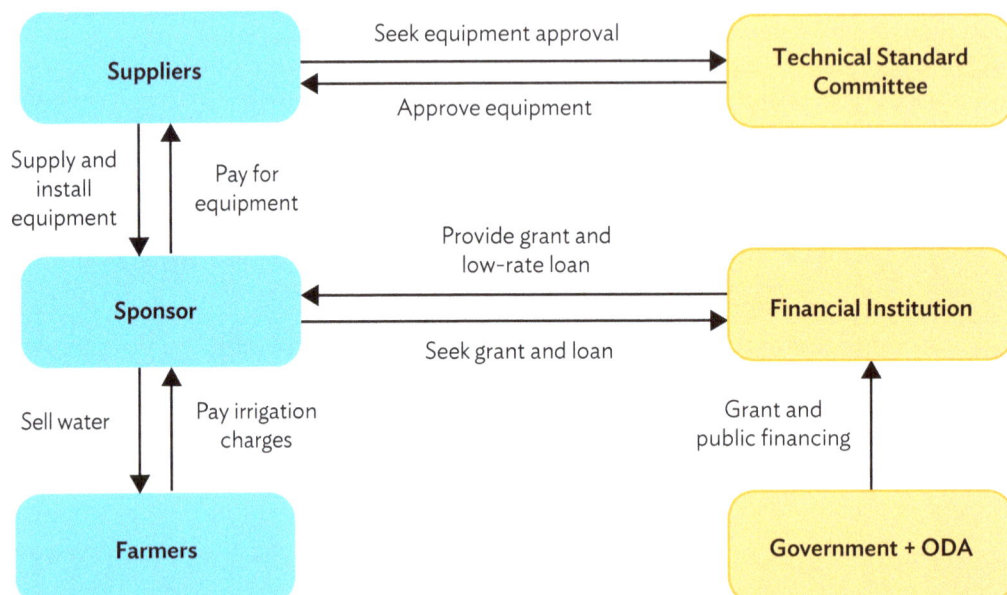

ODA = official development assistance.
Source: Modified and adapted from official presentations by the Infrastructure Development Company Limited (IDCOL).

Under this scheme, professionals manage the water supply and SIP systems, resulting in farmers getting water for irrigation as a service. This saves time and money for farmers, resulting in higher yields and profits. Since an intermediary (the sponsor or agency) is involved, the water price includes the cost of management. However, this reduces the cost to the farmers if it covers large areas.

The fee-for-service modality suits well the larger systems targeting medium-sized farms but not the small farms that complement their subsistence farming with agribusiness activities. This modality can also be implemented to an aggregated group of farms (greater than 5) to provide irrigation services at affordable rates. In such cases, additional financial provisions should be planned for compensating the sponsors for the risk of losses or no profits.

IDCOL has implemented more than 1,500 projects under this model and acquired a large amount of knowledge and experience.

Key Advantages of Fee-for-Service Modality

(i) **Improvement in quality of farmers' lives.** Since the SIP system is entirely operated by the sponsor, farmers no longer need to spend time renting diesel pumps, buying diesel, and looking after the diesel pumps during irrigation. Farmers recognize this as an important improvement in the quality of their lives.

(ii) **Encouragement of professionalization.** Sponsors manage several projects, which helps them to quickly gain experience and take advantage of economies of scale and other synergies. Personnel working for sponsors get specialized in the O&M of SIP systems and quickly become itinerant experts.

Specific Recommendations from Lessons Learned in Pilot Interventions

(i) **Make provisions for mitigating slow business consolidation.** Since the SIP systems are owned by sponsors and not directly by farmers, the transition of farmers using diesel pumps toward cleaner SIP systems is slower than in the ownership model. It takes up to 3–4 years to consolidate enough clients to commence the business for a sponsor. The financing structure for this model should consider specific considerations for mitigating slow consolidation, otherwise payback obligations of the sponsor will make the project fail.

Bangladesh Agriculture Development Corporation (BADC) Solar Irrigation Pump Project. Project funded by BADC in Lalmonirhat under an ownership modality (photo provided by Asian Development Bank consultant).

(ii) **Include support and promote additional agribusiness services.** Financing conditions offered by IDCOL is a grant equal to 50% of the project cost, 15% equity, and a 10-year private sector low-interest loan for the remaining of 35%. Fees charged to farmers cannot be more expensive than what a farmer will spend when using a diesel-operated pump. Under these conditions, and at current equipment prices, the economic feasibility of projects is rather weak. This financing-working model should include additional support and direct promotion for developing agribusiness services, such as machinery renting (for husking, threshing), grinding, and seed distribution, to improve the profitability of a project.

Infrastructure Development Company Limited (IDCOL) Solar Irrigation Pump Project. This 30 kW peak project funded by IDCOL in Dinajpur targeted small and medium-sized farms with subsistence and agribusiness activities.

7.3   Income from Exportable Electricity

Exporting surplus electricity to the grid provides a regular income to farmers and sponsors using an SIP system, reducing the need for additional financial support. This road map estimates that farmers connecting their SIP systems to the power grid could recover up to 80% of their contribution from the sale of their surplus electricity over a 20-year period. Grid connection for projects will depend on the capacity of the nearest distribution network, feeders, distribution transformers, and substations.

The economic feasibility of interconnections depends mostly on the distance to the connecting point and the voltage level of this interconnection. SIP systems with generating capacities of up to 30 kilowatt peak (kWp) and located at short distances to the low-voltage (400-volt) grid could be connected to the low-voltage grid with no problems. SIP systems larger than 30 kWp should go through a technical assessment to determine if they can be better connected to the low-voltage or medium-voltage (11/33 kV) networks. Grid connection costs should be included as part of the SIP project's total costs.

To make the integration of small SIP systems to the grid more manageable, groups of these systems could be aggregated, providing a single communication node with the utility for the group. This reduces problems of attempting to visualize and control the huge number of small generation facilities on the distribution network, because the utility will be able to better interact with a manageable number of distributed generation resources rather than attempting to control each generator individually.

SIP systems produce renewable electricity every day, even when pumps are idle. Pumps operate full time about 100–150 days per year, while the rest of the time solar PV panels keep producing electricity. This provides the opportunity to export electricity to the grid, if connection to the grid for the SIP system is feasible. By the end of the road map rollout, there will be 1,022 MWp installed capacity with the potential to export to the grid 480 GWh per year. The total cumulated exportable electricity over the 8-year road map is 2,390 GWh (Figure 19).

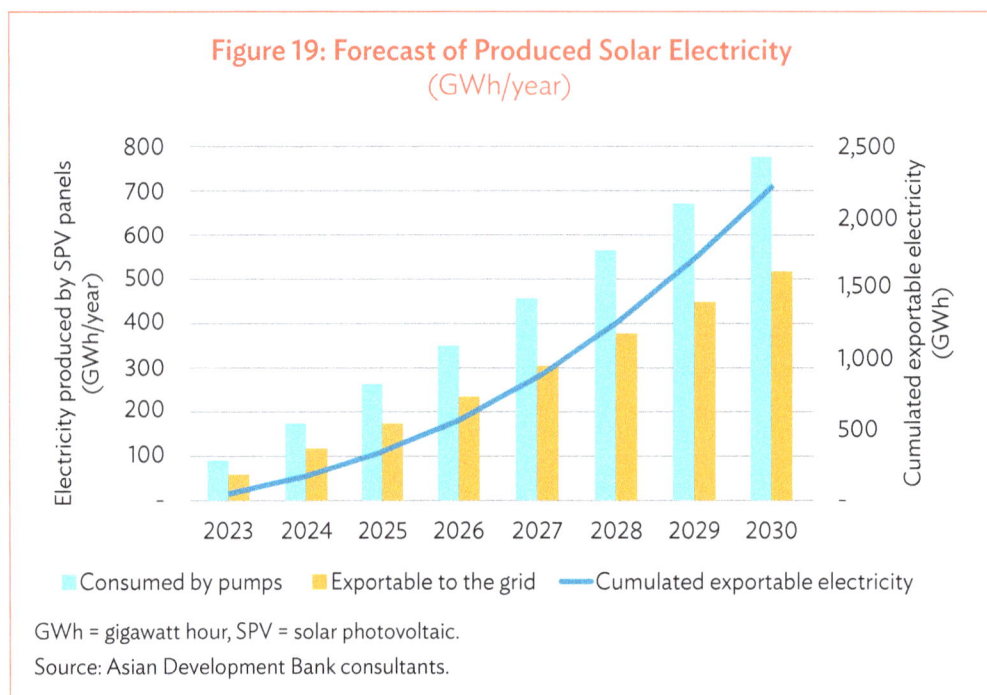

Figure 19: Forecast of Produced Solar Electricity
(GWh/year)

GWh = gigawatt hour, SPV = solar photovoltaic.
Source: Asian Development Bank consultants.

7.3.1   Compensation to "Only Export" Solar Irrigation Pump Systems

This road map recommends establishing power purchasing agreements between utilities and farmers with SIP systems that only export electricity to the grid, i.e., that do not consume electricity from the grid. The road map recommends that for this case, utilities pay the exported electricity at the 33 kV bulk electricity tariff. The latest approved tariffs by BERC range from Tk5.89 to Tk8.24 per kWh (Table 18).[68] The road map further suggests that the Power Division work with BERC to establish an appropriate higher tariff for regions outside of Dhaka, aiming to incentivize farmers to transition to SIPs. It is also recommended that the government set up a mechanism to compensate distribution utilities for additional tariff paid to SIPs. In determining a higher bulk tariff, BERC could consider a rollback to bulk electricity tariffs after the first 5 years, by which time farmers and the project sponsor will have recouped much of their initial investment.

The exported electricity to the grid is valued at $40 million per year of income to farmers and sponsors after the road map has been fully implemented.[69] The maximum total cumulated value of the exportable electricity over 10 years would be $198 million. This could be a good financial option to alleviate the debt acquired by farmers and sponsors installing SIP systems.

Table 18: Bulk Electricity Tariffs at 33 Kilovolts, January 2023
(Tk/kWh)

N°	Electricity Distribution Company/Entity	Bulk Tariff
1	Bangladesh Power Development Board	5.89
2	Dhaka Power Distribution Company Limited	8.22
3	Dhaka Electric Supply Company Limited	8.24
4	West Zone Power Distribution Company Limited	7.12
5	Northern Electricity Supply Company Limited	6.70

kWh = kilowatt-hour, Tk = taka.

Source: Gazette, 30 January 2023. Issued by the Bangladesh Energy Regulatory Commission.

7.3.2   Compensation to "Net-Metered" Dual Solar-Electric Systems

Dual solar-electric systems can operate with solar energy during sunshine hours and with grid electricity during the rest of the day or during the night. While dual solar-electric systems may consume electricity from the grid, they can also export excess renewable electricity to the grid under the net metering policy adopted by the government. When installing a dual solar-electric system, the consumer is required to follow the safety guidelines and technical requirements for interconnection set forth by BERC. A bidirectional meter allows electricity to flow in both directions during the net metering process. The measured data can be sent to a centralized aggregator service or stored in the meter itself.

The net energy recorded on the meter—that is, the total energy extracted from the network less the total energy supplied to the network during the designated billing period—is used to compute the customer's bill. The consumer is responsible for paying the bill for net electricity consumption if the amount of electricity consumed from the grid exceeds the amount supplied to the grid. In contrast, the distribution utility must permit the

---

68   Bulk electricity tariffs were last updated on 30 January 2023 by the BERC.
69   Calculated at an average bulk tariff at 33 kV of Tk7 per kWh.

consumer's entire credit (measured in kWh) to carry over to the following billing cycle if the amount of electricity generated and exported to the grid exceeds the amount supplied by the grid. Any balance of electricity favorable to the consumer at the end of settlement period is paid to the consumer at a specified tariff. Net metering helps to reduce the strain on distribution networks and prevents loss in long-distance transmission and distribution lines. However, *Guidelines for the Grid Integration of Solar Irrigation Pumps 2020* does not foresee the application of the net metering scheme for solar pumps. This guideline would have to be revised to allow net metering in dual solar-electric systems. It is also recommended that bulk electricity and net metering purchase tariffs offered to SIP system owners are frequently revised to ensure a reasonable return on farmers' expenditure and to entice farmers to switch from diesel pumps to SIP systems.

7.4   Guidelines for Project Design

7.4.1   Site Selection

Site selection for the installation of SIP systems can be limited by technical, proprietary, or use constraints. Bangladesh presently has no relevant technical constraints since the terrain is mostly flat across most areas. Proprietary or use constraints should be dealt with on an individual basis.

It is recommended that the rollout of the road map avoids SIP installation on cultivated lands, but installs them instead on dams or canals. To the extent possible, sites should be close to roads to save on transport costs and to a potential grid connection point. It is also recommended that government regulations be enacted to permit grid connections to low-voltage distribution lines to reduce connection costs to farmers. The latter would require BREB and other distribution utilities to strengthen their distribution networks. For the distribution utilities, the government will need to provide funds for the strengthening and adequate upgrade of distribution grids. The SIP system size is strongly linked to demand for water and available water sources. The five steps shown in Figure 20 should be followed for determining the size of the SIP system. An overview of the process for the calculation of pump size is shown in Figure 21.

Figure 20: Steps for Determining the Size of a Solar Irrigation Pump System

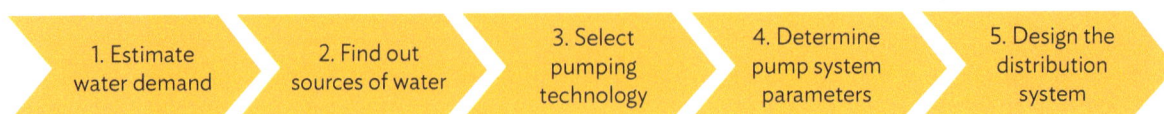

| 1. Estimate water demand | 2. Find out sources of water | 3. Select pumping technology | 4. Determine pump system parameters | 5. Design the distribution system |

Source: Asian Development Bank consultants.

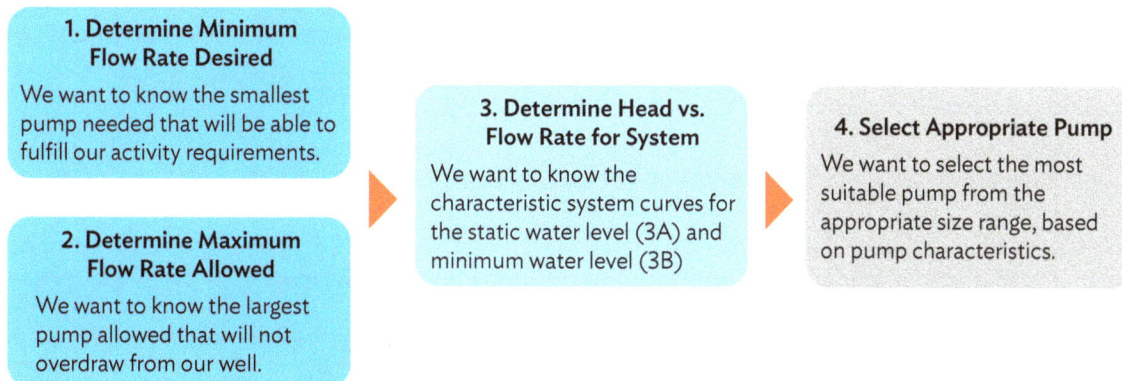

Figure 21: Overview of the Process for the Calculation of Pump Size

1. Determine Minimum Flow Rate Desired

We want to know the smallest pump needed that will be able to fulfill our activity requirements.

2. Determine Maximum Flow Rate Allowed

We want to know the largest pump allowed that will not overdraw from our well.

3. Determine Head vs. Flow Rate for System

We want to know the characteristic system curves for the static water level (3A) and minimum water level (3B)

4. Select Appropriate Pump

We want to select the most suitable pump from the appropriate size range, based on pump characteristics.

Source: Comprehensive Initiative on Technology Evaluation. 2017. *Solar Water Pumps: Technical, Systems, and Business Model Approaches to Evaluation.* Cambridge: Massachusetts Institute of Technology. https://dspace.mit.edu/bitstream/handle/1721.1/115516/Solar%20Water%20Pumps-%20Technical%2C%20Systems%2C%20And%20Business%20Model%20Approaches%20To%20Evaluation%20.pdf?sequence=5.

The land extension needed for a PV module array depends on the SIP system capacity. Usually, 1 kWp PV occupies an average of 13.4 m^2 for a fixed stand. Where possible, it is recommended that PV module arrays are mounted above water storage or water ponds to minimize the use of agricultural terrain. It is also recommended to design the water storage system to be higher than the fields to benefit from gravity.

The configurations shown in Table 19 are recommended for all types of larger size SIP systems.

Photovoltaic module over water storage. A photovoltaic module array of a pump system in the Barind region (photo provided by Asian Development Bank consultant).

Table 19: Technical Configurations for Larger Size Solar Irrigation Pump Systems

Pump Size (kW)	Hydraulic Head (m)	Flux (m³/hr)	PV Capacity (kWp)	Land for Fixed Standard (m²)	Control Inverter (kW)
18.5	9–35	300–80	32.4	2,600	22
22.0	10–25	400–180	37.8	3,025	30
30.0	9–25	600–200	51.3	4,105	37
37.0	8–30	800–200	62.7	5,020	45

kW = kilowatt, kWp = kilowatt peak, m = meter, m^2 = square meter, m³/hr = cubic meter per hour, PV = photovoltaic.

Source: Asian Development Bank consultants estimate.

7.4.2 Estimation of Peak Demand

The first step is to determine the water demand for the project, which very much depends on the irrigation demand of the area to be serviced and the type of crops irrigated. In Bangladesh, most crop cultivation takes place in millions of small family farms that operate on no more than 1 hectare (ha) of land, and in some cases only 0.4 ha. Such smallholders account for 88% of farms and 60% of all cultivated land.

Rice cultivation requires a peak water requirement during land preparation and transplanting. This usually occurs between 10 and 20 January. The Food and Agriculture Organization's climate stations' estimations of effective rainfall have been used to map the peak CWR for Boro rice. The distribution of the peak Boro requirement is highest in the southeast (138 mm per 10 days) and lowest in the northwest (123 mm per 10 days). The design of SIP, especially for individuals, needs to consider the peak CWR.

Irrigation application efficiencies are very variable parameters and depend on soil texture, soil structure, soil deficit at the time of irrigation, and to some extent, discharge of applied water. The irrigation application method is the same as border flooding. Generally, heavy soils can have higher application efficiencies than light-texture soils. Well-structured soils will have higher infiltration rates than lighter soils, which can lead to lower efficiencies. The higher the discharge application, the greater the chance of higher efficiencies.

According to local practices, the peak crop water requirement varies from 37.58 to 12.53 mm per ha, based on different soil types.

The tables for tube well spacing used by *upazila* irrigation committees for approval of new pumps need to be improved by changing the heading of columns from cusecs (cubic feet per second) to cubic meters per day for SIP rollout. This will allow a variation in discharge rates during the day, as affected by solar radiation variations.

7.4.3 Components and Configuration of a Solar Irrigation Pump System

SIP systems are an ideal match for irrigation since water can be pumped when solar irradiation is high. SIP systems are reliable and are powered by clean and affordable energy. In the absence of subsidies, PV electricity can compete with thermal power. SIP systems offer the possibility of storing energy in the form of water by saving water in elevated tanks or reservoirs for later distribution or in battery energy storage systems for use when no sun is available. Battery energy storage system technology is mature but too expensive for SIP applications at present.

SIP systems were first introduced in a few countries about 50 years ago for the provision of drinking water. The technology has since developed around many different designs and applications. It is a very suitable technology for producing drinking water since it can remove dirt and filter sand at low costs. SIP systems can also be combined with other technologies such as reverse osmosis, ultraviolet sterilization, or ozone sterilization to remove bacteria and salinity in water. They can also be combined with ion-exchange resins or other filters to remove arsenic and other heavy metals to purify contaminated water. The classical configuration of SIP systems for irrigation with groundwater is shown in Figure 22.

It is logical to connect SIP systems to the grid where economically feasible. Grid-connected SIP systems can export clean (solar) energy to the grid when its pumps are not operating, especially in seasons with low irrigation demand. Grid-connected SIP systems can use grid electricity during cloudy weather to avoid irrigation shortages.

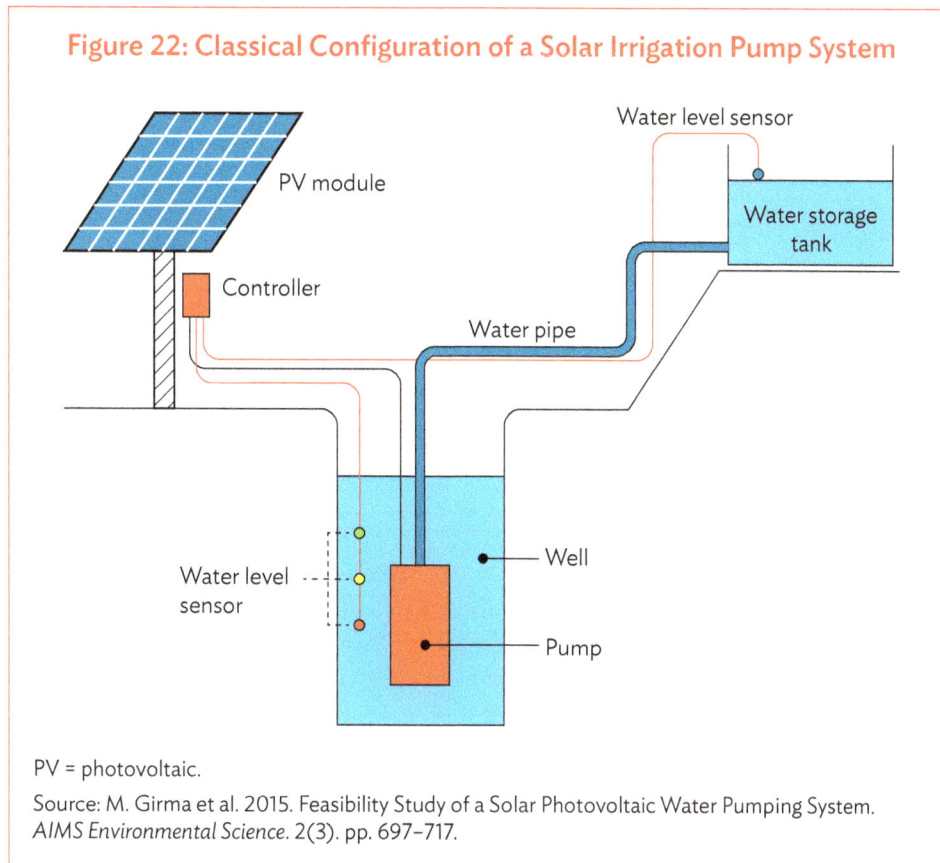

Figure 22: Classical Configuration of a Solar Irrigation Pump System

PV = photovoltaic.
Source: M. Girma et al. 2015. Feasibility Study of a Solar Photovoltaic Water Pumping System. *AIMS Environmental Science.* 2(3). pp. 697–717.

An SIP system is a mature technology and is further improving. An SIP system consists of

- PV modules;
- electronic components, cabling, and protection;
- a pump set;
- a storage tank and water distribution system; and
- a control room.

It is easy to install SP systems, and they require minimum operational input. Their discharge rates are low, but they pump over a long period (8-10 hours a day). Hence, they will cause minimum interference with neighbouring wells. Conventional submersible pumps last for 5–8 years, and conventional centrifugal pumps last for more than 30 years; good-quality PV panels can last more than 25 years and modern inverters can last for about 10 years.

Both the reliability and maintenance requirements for SIP systems have improved since initial systems were introduced. Maintenance is required every 1–5 years, while PV modules only need to be kept clean of dust. Inverters are the most maintenance-intensive part of these systems and require careful attention to ensure they operate properly.

Several thousand SIP systems have been installed internationally. Bangladesh should consider the lessons learned from installations to date, including the following:

- **The quality of PV modules is important.** There are many manufacturers and qualities of PV modules available in the market, and the efficiency and especially the lifetime of PV modules also vary largely. Only certified PV modules that have passed proper tests using international standard IEC 61215-1 for "Terrestrial photovoltaic (PV) modules–Design qualification and type approval" should be considered.[70]
- **Sunshine hours and angles must be correctly calculated.** The correct number of sunshine hours must be determined before estimating the right size/capacity of systems. Alignment of the PV modules at the right angle to receive maximum sunshine throughout the day is also important for achieving the maximum electricity output. An overestimation of sunshine hours or lack of tracking system to adjust the angle of PV modules will reduce drastically the performance of the system.
- **Sophisticated control technology requires available technical assistance.** Electronic control systems are sophisticated and in many cases limited technical services are available to solve problems in these systems. SIP programs must consider this risk and make sure that proper training is given to operators and users, and that good quality technical service is available within a feasible distance of the installations.
- **Response capacity should be considered.** The inability of SIP systems to supply immediate variations in water demand is considered a weakness, especially in places with very frequent severe weather events during the irrigation period. In such places, the irrigation systems should include either a genset or a connection to a grid as backup to operate the pump.
- **A good quality inverter is crucial.** The inverter must include good quality maximum power point tracking (MPPT) to optimize the match between the PV module and the pump, the battery bank, or utility grid, a frequency inverter IP 65, a surge protector, a combiner box, and remote monitoring to keep the efficiency of the inverter at optimal levels. This will ensure the long-term, efficient operation of the systems.
- **Efficiency losses should be considered.** Other losses of efficiency must be considered for the optimal operation of SIP systems. Among them are the losses caused by high temperatures, dirt over PV panels, core and copper loss in the electrical installations, and pipe resistance.
- **Motor cooling is important.** The cooling system of motor driving pumps is important and should be selected according to operation parameters and pumped water properties (temperature, sediments, and presence of sand).
- **Resistance to extreme weather conditions is important.** Systems and PV modules must be installed to resist extreme weather conditions at specific locations, e.g., high wind speeds. Locations that may suffer level 4+ hurricane speed (wind speeds faster than 200 km per hour) need very special mounting structures, and flood-prone areas need high standers.
- **Land area needed should be considered.** Configurations using DTWs and LLPs are used for irrigating large areas that require large amounts of water, and therefore require more energy and consequently more area for the PV modules. For these configurations, design should consider using the land under the modules or build the PV array over water to avoid occupying large extensions of agricultural land. Several examples exist of using the land underneath the PV modules for growing vegetables that are not tall and support some shadowing, or for poultry or fish farming.

---

[70] The requirements for the type approval and design qualification of terrestrial photovoltaic modules that are suitable for long-term operation in general open-air climates—as defined by IEC 60721-2-1—are outlined in the standard IEC 61215-1. This standard is meant to cover all materials for terrestrial flat plate modules, including thin-film and crystalline silicon modules. This test sequence aims to ascertain the electrical and thermal properties of the module and demonstrate, to the greatest extent feasible given reasonable time and cost constraints, that the module can withstand extended exposure in the climates covered by the scope.

7.4.4   Selection of Pump

Since farmers have practically always used some sort of pump, they are the ones who understand the system's pumps the best. Most SIP systems will use surface or submersible centrifugal pumps. The surface centrifugal pumps are usually referred to as centrifugal pumps, while submersible centrifugal pumps are referred to as submersible pumps.

Centrifugal pumps are suitable for relatively low hydraulic heads and large water flux. Submersible pumps are installed in the bottom of the borehole, which has a diameter of 12–20 cm, using specialized equipment. Submersible pumps are electrically wired for feeding and control through a pipe to the surface while submerged in water. Submersible pumps are much easier to start than surface pumps, which require special start-up equipment to be added. Submersible pumps are made of stainless steel or iron casting. Pumps made of stainless steel lift cleaner water since they do not produce any rust. Pumps made of iron casting however have a longer lifetime. Submersible pumps have a shorter lifetime compared to surface pumps. However, sewage centrifugal pumps offer an effective solution for gathering contaminated sewage and sandy water. By using centrifugal force, these pumps generate flow and pressure, enabling them to transport liquids over long distances or against elevated water pressures.

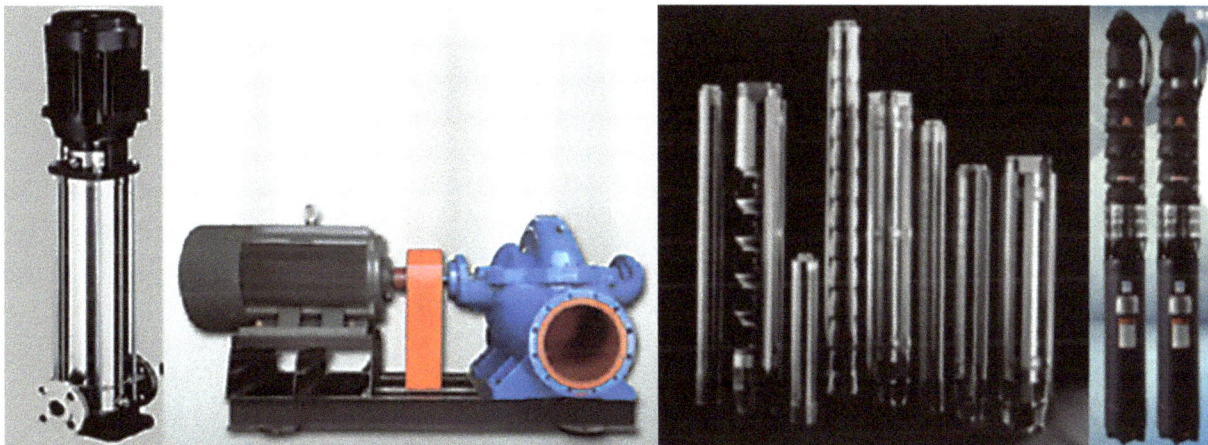

Centrifugal surface pump (photo provided by Asian Development Bank consultant).

Submersible pumps (photo provided by Asian Development Bank consultant).

The pump rating or pump power is calculated as the product of the water demand (flux) and the total water head that the pump must provide. Power refers to the electric power supply the pump needs for normal operation. Flux is the water volume that the pump can lift in a unit time, usually expressed in volume per hour. The head is the equivalent height the pump must lift the water. The total head results from several contributions as shown in Figure 23 and the formula below.

$$H_T = H_G + H_S + H_D + H_F$$

where

H_T Total head of the pump

H_G Head of the outlet pipe above the ground (assuming the outlet pressure is negligible)

H_S Static head due to the depth of the water level in the well, in absence of any pumping

HD   Dynamic "drawdown" head, or the effective water level that is dynamically lowered by the water flow extraction

HF   Friction losses in the piping circuit, depends on the flow rate and pipe type

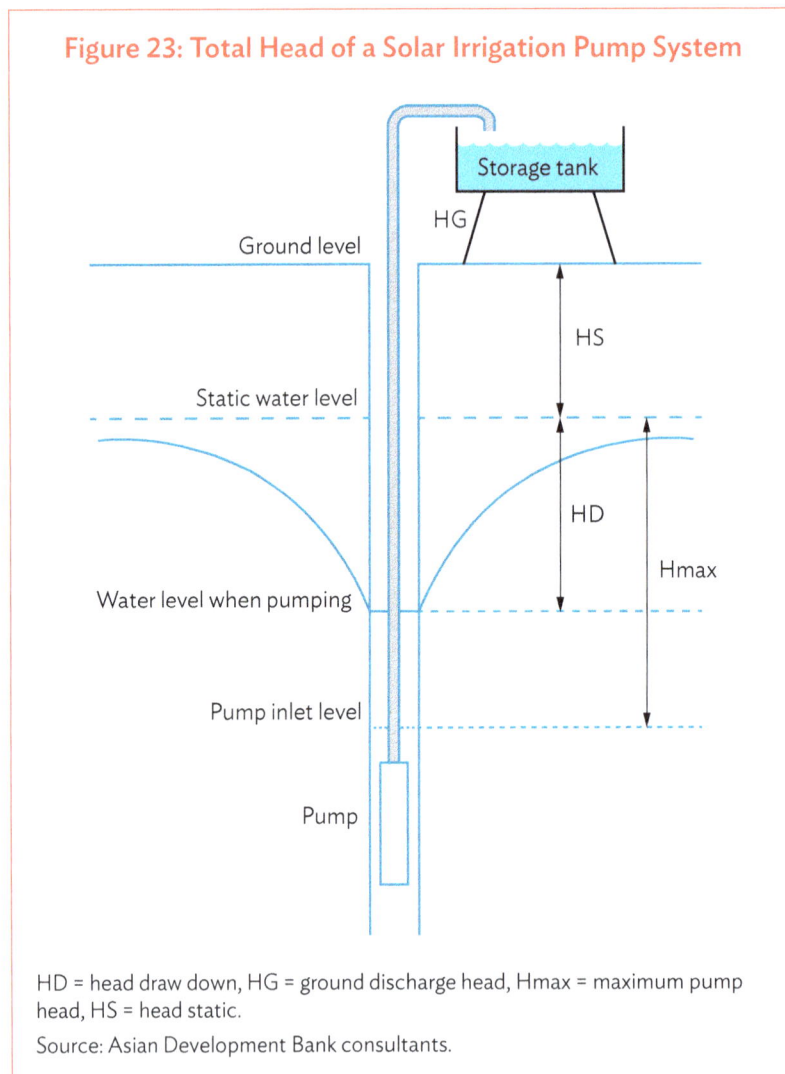

Figure 23: Total Head of a Solar Irrigation Pump System

HD = head draw down, HG = ground discharge head, Hmax = maximum pump head, HS = head static.

Source: Asian Development Bank consultants.

SIPs are moved by a motor. Motor types include alternating current (AC) and direct current (DC). These motors can be synchronized with three-phase permanent magnets or have three-phase induction. They may be with brushes or without brushes. In SIP systems, DC motors are easier to control than AC motors, since PV modules produce DC electricity. However, DC motors are more expensive than AC motors and usually small (less than 2.2 kW). AC motors are a more suitable solution for most SIP systems. AC motors require an inverter to convert DC electricity generated by the PV modules to AC.

Hybridization of electrical pumps. Electrical pumps can easily be turned into SIPs, becoming a hybrid solar grid. This is a great advantage because SIP systems can be connected to the grid and operate with both solar energy and grid electricity with no technical issues.

Hybridization of diesel pumps. Diesel pumps cannot be easily converted into SIPs. They need a special axis connection and a switch to change back and forth from diesel to solar. This type of hybridization is not recommended.

7.4.5 Water Distribution

To reduce wastage and evaporation losses, the water distribution system should be placed underground. Unplasticized polyvinyl chloride pipes of class B are best suited to this purpose.

If the SIP system will also be used for producing drinking water, it is recommended to install drinking water containers for storing water needed for at least 5 days.

7.4.6 Solar System

PV modules are made of PV panels that generate electricity from the sun's light. The solar pump system's most costly part is without a doubt the solar panels. Since the power required for the pump determines the size of the array, even a slight variation in the pump's power can have a significant effect on the system's total cost. To guarantee that the panels receive the greatest amount of direct sunlight possible in the morning and afternoon, they can be manually tracked in a single axis or fixed. Depending on how much sunlight the PV modules receive throughout the day, their power output will change. PV modules are linked in series, and occasionally long strings of modules are linked in parallel to create expansive arrays. Typically, SIP systems are powered by either AC or DC that is generated directly from an array of PV modules. Three types of solar panels are available today:

- monocrystalline panels;
- polycrystalline panels; and
- thin-film amorphous panels.

Monocrystalline panels are slightly more efficient than polycrystalline ones. However, monocrystalline panels are more expensive, and therefore are used less. Polycrystalline PV panels of 270 watt-peak (Wp) with 13%–16% efficiency is the most popular and widely used in large-scale PV stations and most suitable for SIP systems as well.[71] However, in recent years, there has been a shift toward using 535–600 Wp panels that have a higher panel efficiency (21%). Thin-film panels are not recommended because of their low reliability and cost.

The PV modules should be mounted facing south at an optimized tilt of a certain angle (degree) to maximize energy yields throughout the year. As the irrigation season in Bangladesh is mostly from January to March, the PV tilt angle should be 25° (local latitude). For installations where solar modules are permanently mounted, they should be tilted for maximum output based on previous experiences and available technical studies. As a rule, the power output is optimized for meeting peak irrigation demand. To reduce voltage loss in the system wiring, it is suggested that PV modules should be installed as close as possible to the water source. A fence around the PV modules is required to protect the modules from theft and damage. After installation, the area inside the fence must be maintained on a timely and routine basis. The fence could be constructed with concrete poles and galvanized barbed wire. Firm structures such as concrete bases are necessary for long-term durability and public safety. PV mounting structures can be locally manufactured with hot galvanized iron. Considering flooding risks, the design of structure heights should minimize potential damage.

---

[71] Commonly used panels range from 250 to 365 Wp. They are manufactured by LG, Sunpower, and Hanwha Q cell.

Ratio PV Module–Pump

The optimal ratio of PV modules and pump rating (P_{PV}/P_{pump}) depends mainly on the amount of solar irradiation and the tilted angle of the PV module array, but also on other factors such as losses due to temperature in the panel cells, dust, losses in the inverter, and wiring.

The optimal ratio P_{PV}/P_{pump} can be calculated by this formula:

$$P_{PV}/P_{pump} > 1 + C_{unused\ irrad} + C_{decay\ PV} + (T_{max\ PV} - T_{ref}) * C_{c\text{-}Si} + C_{dust\ loss} + C_{inverter\ loss} + C_{wire\ loss} + C_{other\ losses}$$

where

$C_{unused\ irrad}$	Coefficient of unused irradiance. This coefficient is set to 20% for pumps operating at full capacity for 5 hours per day. A coefficient of 30% should be used for pumps operating at full capacity for 6 hours per day.
$C_{decay\ PV}$	Coefficient of decay of PV modules for 25 years. This coefficient is set to 20%.
$C_{c\text{-}Si}$	The crystalline silicon (c-Si) PV module power coefficient is set to 0.4%/°C.
$C_{dust\ loss}$	Coefficient of losses due to dust. This coefficient is set to 5% for medium-range cover ratio.
$C_{inverter\ loss}$	Coefficient of losses in the inverter. These losses usually account for 5%.
$C_{wire\ loss}$	Coefficient of wire losses. These losses usually account for 3%.
$C_{other\ losses}$	Coefficient for other losses, including measurement errors at the PV module, set to 2%.
$T_{max\ PV}$	Maximum temperature that cells of PV panel can reach during a hot day. For Bangladesh, in a day with an environment temperature of 40°C, the temperature of the solar cells in the module could reach 70°C.
T_{ref}	Temperature reference set to 25°C.

Estimation of $C_{unused\ irrad}$

The solar irradiance distribution follows a sinusoidal law as seen in Figure 24. Three hours of full power from the PV modules will result in three hours of the pump operating at full capacity. This should be around noon time, starting at 72.5° solar hour angle, which means solar irradiance over 923.7 W per m² between 10:30 a.m. and 1:30 p.m. If two more hours of pumps working at full power is needed, starting at 9:30 a.m. and finishing at 2:30 p.m., then it will reach the 20% standard test conditions solar irradiance losses. If pumps are estimated to work at full capacity for 6 hours per day (from 9:00 a.m. to 3:00 p.m.), then the standard test condition solar irradiance losses would account for 30%. The pump will start running at lower than its maximum capacity when solar irradiance reaches 30%, at about 7:30 a.m., and will continue running also at lower than its maximum capacity until 4:30 p.m.

Time of solar irradiance varies with seasons and weather. The amount of solar energy received per day and per unit by season and latitude in kWh per m² is shown in Table 20 (latitude for Bangladesh ranges from 20.86382° to 26.33338°). For illustration of this variation, Figure 25 shows the daily standard sunshine hours per month of the year.

Figure 24: Solar Irradiance Distribution during Daytime
(kW/m^2)

kW/m^2 = kilowatt per square meter.
Source: Asian Development Bank consultants.

Figure 25: Standard Daily Dhaka Sunshine per Month of the Year
(hours)

Source: M. S. I. Sayeed. 2020. Sustainable Solar Water Pumping for Irrigation in Bangladesh. *ISES Conference Proceedings*. http://proceedings.ises.org/paper/solar2020/solar2020-0016-Foster.pdf.

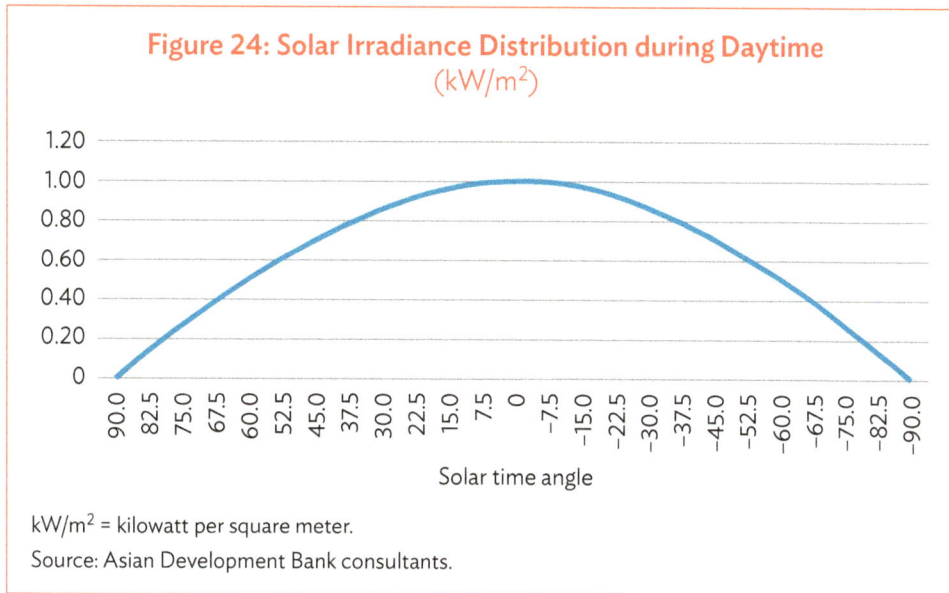

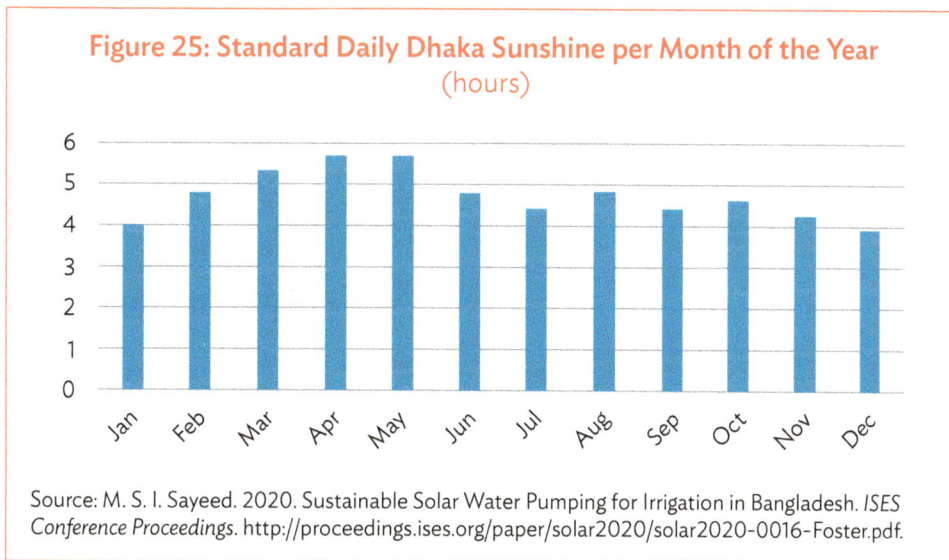

Table 20: Amount of Solar Energy Received per Day by Season and Latitude
(kWh/m^2)

Latitude (deg)	Summer Solstice	Equinoxes	Winter Solstice	Average
Solar constant	1.321 kW/m^2	1.366 kW/m^2	1.412 kW/m^2	
90	12.64	0	0	3.1600
80	12.45	1.8	0	4.0125
70	11.88	3.55	0	4.7450
60	11.49	5.19	0.58	5.6125
50	11.59	6.67	2.04	6.7425

continued on next page

Table 20 *continued*

Latitude (deg)	Summer Solstice	Equinoxes	Winter Solstice	Average
40	11.65	7.95	3.76	7.8275
30	11.39	8.99	5.42	8.6975
20	10.99	9.75	7.12	9.4025
10	10.25	10.22	8.61	9.8250
0	9.20	10.38	9.84	9.9500
-10	8.06	10.22	10.96	9.8650
-20	6.67	9.75	11.75	9.4800
-30	5.07	8.99	12.18	8.8075
-40	3.52	7.95	12.45	7.9675
-50	1.91	6.67	12.39	6.9100
-60	0.54	5.19	12.28	5.8000
-70	0	3.55	12.70	4.9500
-80	0	1.80	13.31	4.2275
-90	0	0	13.51	3.3775

deg = degree, kW/m^2 = kilowatt per square meter.

Source: Asian Development Bank consultants' estimate using formulas for the zenith angle of the sun at each hour of the day.

For Bangladesh, a ratio P_{PV}/P_{pump} > 1.73 for pumps operating 5 hours per day at full capacity and >1.83 for pumps operating 6 hours per day at full capacity. Given that winter is Bangladesh's most crucial irrigation season, a reasonable ratio is set at 2. It is important that P_{pump} is the actual power supplied by the pump, which is less than its rated power.

The Importance of PV Modules Tracking

The Rabi crops are sown around mid-November after the monsoon rains are over, and harvesting begins in April or May. The irrigation for these crops occurs in the period of lowest sun irradiance (December–March). As optimal tracking of the PV modules helps to increase water output by at least 20%, an automated or a mechanical tracking system is important. Mechanical tracking systems are a good low-cost choice for places with operators that can take care of them during operation hours. SIP systems using manual tracking standers can pump about 20% more water than systems using fixed standers. Manual tracking standers require that someone turns the panels every 1–2 hours. Manual tracking standers, in particular double-axis tracking standers, suffer with strong wind, so only manual single-axis tracking standers are recommended for SIP systems in Bangladesh. Double-axis tracking standers are also not recommended because of their high failure rates and higher costs.

Controllers, Cabling, and Protection

Although solar panels generate DC power, the majority of pumps used in agriculture are AC pumps. The electronics transform that DC power into AC so that it can be utilized with the pumps. They are often kept in a weatherproof box beneath the panels. The electronics box typically includes an on/off switch as well.

Controllers and inverters, also called variable speed drives, regulate the operation of SIP systems and are their most important component. Their basic functionalities include the control of starting and stopping the SIP system as well as solar irradiance, MPPT, voltage adjusting, lightning protection, no-load protection, load smoothing, and automatic operation.

Single-axis tracking. Solar irrigation pump systems using manual single-axis tracking standers (photos provided by Asian Development Bank consultant).

Double-axis tracking. Solar irrigation pump systems using manual double-axis tracking standers (photo provided by Asian Development Bank consultant).

Additional features of controllers include remote data collection and monitoring and reporting to the controller. The communication between the SIP systems and the monitoring center is usually done through mobile phone networks. This enables data collection and potential centralized O&M services.

Although they can be integrated into the submersible motor-pump set in certain SIP systems, controllers are typically mounted on the exterior. DC controllers typically operate with a fixed voltage setpoint and are based on DC to DC control. The PV array's DC power is converted to AC power by AC controllers, or inverters, frequently using MPPT.

Connection cables for all modules are 4 mm^2 PV cables, according to regular practice for SIP systems. The voltage loss should remain within 2% and PVC pipes should be used for all underground cables. Lightning protection should be installed on top of the control room.

Control Room

Controllers and inverters should be placed in the control room or powerhouse. The typical size for a control room is about 7.5 (3 x 2.5) m^2 with a height of 3 m. The controllers should be installed at a higher level than the ground to prevent damage from potential flooding.

7.4.7 Connection to the Grid

SIP systems can be connected to the grid individually or grouped (Figures 26 and 27). Grid integration is a critical success factor of the SIPs to sustain. The road map recommends separate investment projects for grid integration of existing solar pumps that are already installed. Distribution utilities, such as BREB, may undertake such projects.

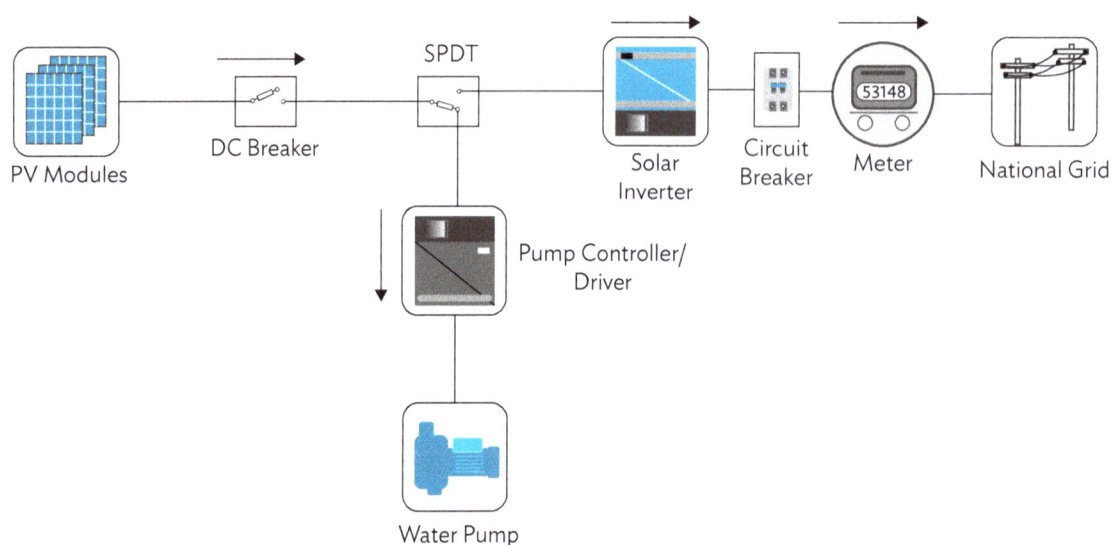

Figure 26: Grid Connection of an Individual Solar Irrigation Pump System

DC = direct current, PV = photovoltaic, SPDT = single-pole double throw.

Source: Government of Bangladesh, Ministry of Power, Energy and Mineral Resources. 2020. *Guidelines for the Grid Integration of Solar Irrigation Pumps 2020*. Dhaka.

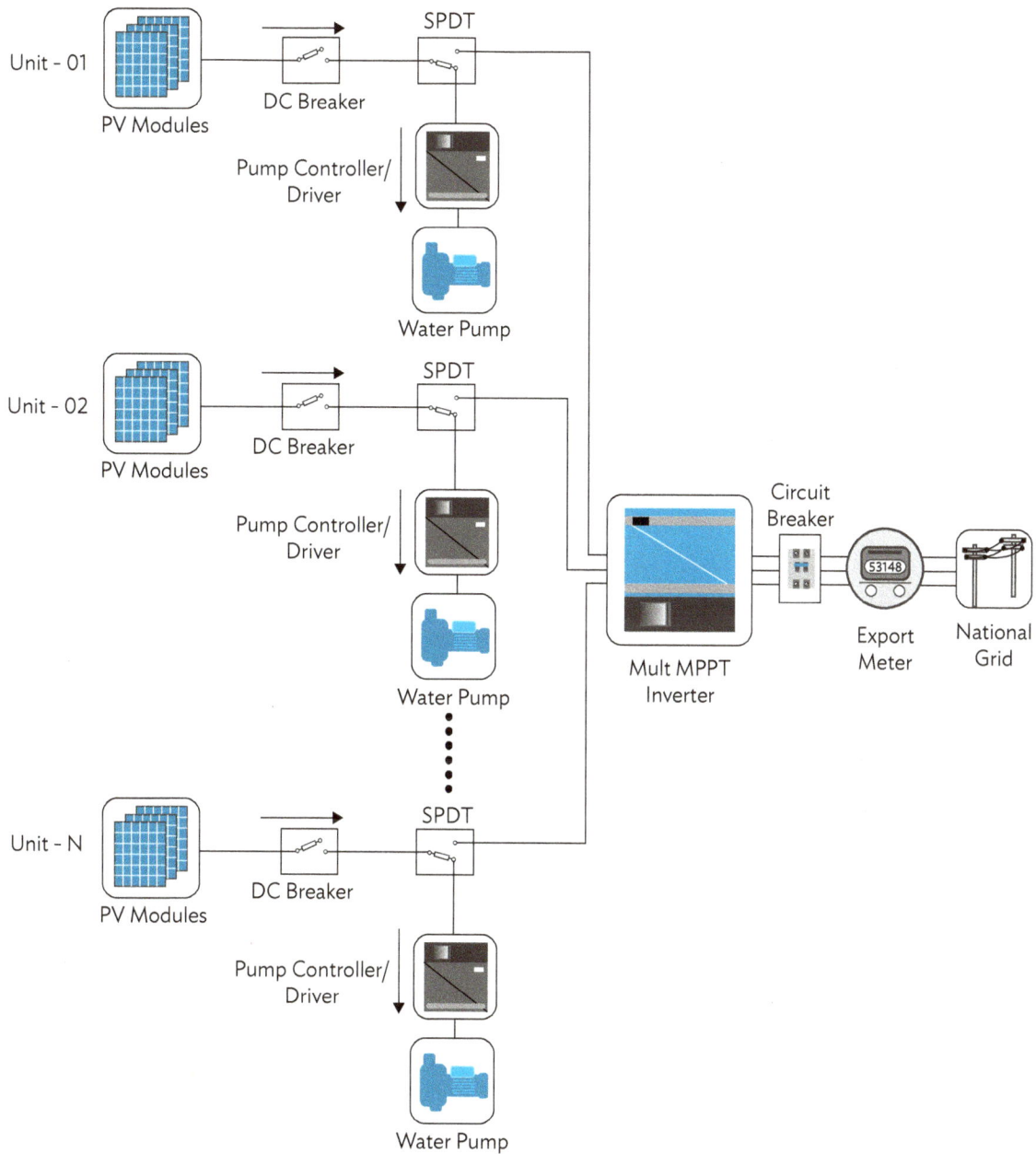

**Figure 27: A Grid Connection of Multiple Solar Irrigation Pump Systems
with a Multi Maximum Power Point Tracking Inverter**

DC = direct current, MPPT = maximum power point tracking, PV = photovoltaic, SPDT = single-pole double throw.

Source: Government of Bangladesh, Ministry of Power, Energy and Mineral Resources. 2020. *Guidelines for the Grid Integration of Solar Irrigation Pumps—2020*. Dhaka.

8. Road Map Rollout

8.1 Sizes and Costs of SIP Systems

8.1.1 Size of Projects

The most relevant factor for determining the size of an SIP system is the number and size of farms the specific project is serving. Irrigation committees usually comprise about 50 farmers, but it is the average size of the farms integrating that will determine the size of the SIP system project. An indicative size of the SIP systems proposed to be installed under this road map is shown in Table 21, together with the estimated irrigated area, number of farmers benefited, and diesel pumps replaced, based on IDCOL's experience installing more than 1,500 SIP systems.

Table 21: Estimated Irrigated Area, Number of Benefited Farmers, and Diesel Pumps Replaced per Different Sizes of Solar Shallow Tube Wells

Pump Rating (kW)	Irrigated Area (ha)	Farmers Benefited	Smaller Diesel Pumps Replaced
5.5	5	16	3
7.5	7	22	4
11.0	10	29	5
15.0	12	36	6
18.5	16	48	8

kW = kilowatt, ha = hectare.

Source: Infrastructure Development Company Limited information provided in October 2020.

8.1.2 Cost of an SIP System

Traditionally, SIP systems were costly largely due to the high cost of their components (pump, PV modules, control systems, and the associated pumping and storage infrastructure). However, technology developments over the years have now made them economically feasible. SIP systems are modular and can be upgraded over time. Components are standard equipment that can be shipped by regular shipping methods available all over the world. However, the cost of SIP systems may vary from district to district, and to some extent it will be a function of the boring length and the length and type of water distribution system involved. Therefore, the specific cost of a project can only be determined when a project area and purpose are identified.

Prices of solar pumps have decreased over the years. Solar pumps nowadays have high power output and low maintenance, and are efficient and commercially available. They are usually made of stainless steel to ensure proper sanitation and long life while submerged under water. The stainless steel resists corrosion when suspended in water for years and is much better than other materials used to manufacture conventional pumps. Brushless pumps have higher electromechanical efficiencies. One of their main benefits is not requiring removal from the well to change the brushes, like in other motors. Brushless motors with MPPT controllers are even more efficient than brushed motors. The cost of pumps may be affected by the pumping head required.[72] The costs of water volume per unit are proportional to the pumping head.

PV panels and their controllers are the largest costs in SIP systems. Prices of PV panels have fallen rapidly in the past 15 years, while their efficiency has been improving. The capacity requirements and thus the PV module size is determined by the intensity of sunshine and number of sunshine hours. The ratio of PV module to pump rating considerably influences the upfront costs of the system. This ratio has been estimated as 2 for Bangladesh (section 7.4.7). A larger ratio means a larger capital cost of the system. Improvements in solar cell manufacturing continues and it is expected that prices of PV modules will keep coming down in next few years. Solar panels are reliable and do not require any maintenance, besides a quick wash to clean the panels of dust and dirt that can affect its operational efficiencies.[73]

Different sizes of SIP systems can be used for minor irrigation in Bangladesh, with pump ratings ranging from 4 to 25 kW, and an average PV module to pump ratio of 2. The International Solar Alliance carried out in 2019 a price discovery bidding for 270,000 SIP systems of sizes ranging from 1 horsepower to 10 horsepower for 22 of their members.[74] The results of this price discovery bidding were a large range of prices. Table 22 shows the minimum and maximum prices recorded for the supply of SIP systems installed, tested, and commissioned. These prices do not include costs for water storage infrastructure and water distribution. The large variation in prices is explained mostly by a variation of quality, guarantee, after-sales service, and the higher cost that well-known brand names with good reputations charge for their products. This large variation of prices also represents the dynamism of the current solar pump market.

Table 22: Bid Prices of Solar Irrigation Pump Systems Recorded in the International Solar Alliance Tender, 2019
($)

Pump Category	Rating (in horsepower)	Lowest and Highest Price[a]
AC Surface Agricultural Pump	1	2,058–18,273
	2	2,405–19,019
	3	3,263–20,671
	5	4,205–28,281
	7.5	6,024–32,065
	10	7,803–36,289

continued on next page

[72] The pumping head is the distance over which the water needs to be moved.
[73] Depending on the location, solar panels may require cleaning either monthly or every year or two.
[74] Bids from five multinational companies were presented in September 2019. Members benefiting from this tender were the following: Benin, Cabo Verde, Congo, Djibouti, Fiji, Guyana, Mali, Mauritius, Nauru, Niger, Peru, Senegal, Somalia, Sudan, South Sudan, Sri Lanka, Tonga, Tuvalu, Togo, Uganda, Yemen, and Zambia.

Table 22 *continued*

Pump Category	Rating (in horsepower)	Lowest and Highest Price[a]
AC Submersible Pump	1	1,881–17,618
	2	2,230–18,096
	3	3,111–20,188
	5	4,166–28,499
	7.5	6,105–32,109
	10	7,905–36,654
DC Surface Agricultural Pump	1	2,216–18,273
	2	2,634–19,019
	3	3,524–20,671
	5	4,514–28,739
	7.5	6,029–32,065
	10	8,202–36,451
DC Submersible Pump	1	2,023–17,618
	2	2,312–18,096
	3	3,183–19,846
	5	4,227–28,022
	7.5	6,468–31,170
	10	8,328–35,061

AC = alternating current, DC = direct current.

[a] Price includes supply, customs clearance, local transportation, installation, and testing and commissioning of complete system and services.

Source: *Bid Report for International Competitive Bidding (ICB) Conducted for Approx. 272,579 Nos. Solar Water Pumping System for 22 Member Countries of International Solar Alliance (ISA)*. https://isolaralliance.org/uploads/docs/4366a51c0d898bc51b607e1af0f840.pdf.

Based on all these assumptions, an estimation of average costs for systems with various pump ratings is presented in Table 23. The variation in prices is mostly dependent on the head of the pump, depth of well, and civil works specifically needed for water storage and distribution. An SIP system cost for Bangladesh would be $3,300 to $4,700 per kW pump installed, including underground water distribution lines and grid connection costs. In terms of land to be irrigated, these costs amount to $330–$470 per irrigated bigha.[75]

The global price of solar PV projects in recent years has shown a downward trend that may continue for the next decade; the price of SIP projects in Bangladesh have gone down 40%–60% in the last 2 decades. The price reduction of PV panels and inverters is expected to slow in coming years, but they will also achieve higher efficiencies. On the other hand, the cost of civil works in Bangladesh has risen over the past 5 years. The combined effect may result in an overall project cost reduction of up to 15% until 2030 compared to current prices.[76]

---

[75] Bigha is a traditional unit of land in Bangladesh, with land purchases still being undertaken in this unit. One bigha is equal to 1337.8 square meters.
[76] ADB consultants' estimate.

Table 23: Estimated Average Costs for Solar Irrigation Pump Systems in Bangladesh ($)

Costs	Pump Rating (kW)				
	5.5	7.5	11.0	15.0	18.5
Equipment[a] (pump, PV modules, controllers) Installation[b]	15,000	22,000	29,000	37,000	40,000
Water distribution lines[c]	6,000	7,000	10,000	14,000	16,000
Grid connection[d]	5,000	5,000	6,000	6,000	6,000
Total cost with grid connection[e]	**26,000**	**34,000**	**45,000**	**57,000**	**62,000**

kW = kilowatt, PV = photovoltaic.

[a] Equipment on site, includes applicable taxes and warranties.

[b] Includes consultancy service, technical assistance, mounting structure, foundations, wiring, and testing.

[c] Estimated length of buried pipes: 5.5 kW = 700 m; 7.5 kW = 800 m; 11 kW = 1000 m; 15 kW = 1200 m; 18.5 kW = 1500 m. Includes materials and civil works.

[d] Includes grid-tied inverter, energy meter, cable, etc.

[e] Grid-connected solar irrigation pump system including irrigation infrastructure.

Source: ADB consultants estimate.

In addition to SIP system costs, cost of hardware, software, and communication network devices for remote monitoring should be considered. Such costs could come from the service charges allocated to implementing agencies. In any case, the allocation of these costs should be carefully considered during the detailed planning stages of the SIP rollout.

8.2   Road Map Target

8.2.1   Number of SIP Systems

The road map estimates that with an efficient use of water resources and with a better organization of irrigation schedules and services, the existing number of diesel-operated pumps in Bangladesh could be replaced by about 300,000 SIP systems. For the period 2023–2031, the road map proposes the installation of up to 45,000 SIP systems, either pure solar or dual solar-electric systems operating with solar and grid electricity. The targeted 45,000 SIP systems will amount to an estimated 1 GWp new solar capacity connected within the country.

The rollout of the road map should prioritize surface water irrigation with LLP and tube well irrigation, operated by agencies such as the BWDB, BMDA, and BADC. For projects supported by the BWDB and BADC, a group of farmers owning the SIP system to collectively irrigate their fields may be a good option to promote.

Since diesel LLP irrigating with surface water accounts for almost 14% of all diesel pumps, and they irrigate 34% of the land, it is proposed that a target should be one-third of the total SIP systems being solar LLPs (i.e., 15,000 SIP systems). The other 30,000 SIP systems would be tube wells. Considering that about 2,200 diesel DTWs are operated by the BADC and BWDB, and could be replaced with SIP systems, the road map proposes that at least 2,000 solar DTWs are installed during the road map period, and the remainder are solar STWs (Table 24).

Table 24: Indicative Target per Type of Solar Irrigation Pump System

Type of SIP System	Indicative Target (units)
Low-Lifting Pumps (LLPs)—Surface water	15,000
Deep Tube Wells (DTWs)—Groundwater	2,000
Shallow Tube Wells (STWs)—Groundwater	28,000

SIP = solar irrigation pump.

Source: Asian Development Bank consultants.

Implementing agencies must design action plans at district level. Prior to the design, each project area should be surveyed in detail. The first step is for all involved agencies to refine the selection of *upazilas* where SIP systems will be installed by using nonphysical criteria, such as farmer preferences. Agencies should define the time and location for SIP system installations and develop action plans that build on the experiences gained in all previous pilots related to SIP systems. The cost of irrigation using installed SIP systems in these action plans should not exceed the farmers' existing diesel-operated pump's irrigation cost.

The targeted 45,000 SIP systems may replace up to 200,000 smaller diesel pumps to irrigate about 400,000 ha of land and serve more than 1.3 million farmers (Table 25). If implemented, these SIP systems would displace the consumption of 300,000 tons of diesel fuel annually or an estimated $380 million in annual savings to the government. Since the government's preference is to use surface water for irrigation, then financing should be prioritized to projects targeting irrigation with surface water. Motorized pumps in need of replacement should be offered the possibility of switching to dual solar–electric SIP systems, which may contribute to peak shaving of the main grid.

Table 25: Targeted Road Map Solar Irrigation Pump Systems and Their Impacts

Type of SIP System	Pump Rating (kW)	Number of Systems	PV Capacity (MWp)	Irrigated Area (ha)	Benefited Farmers	Target Number of Diesel Pumps to Be Replaced
Small size	< 8 kW	18,000	200	100,000	330,000	50,000
Medium size	> 8 kW	27,000	800	330,000	1,000,000	150,000
Total		45,000	1,000	430,000	1,330,000	200,000

ha = hectare, kW = kilowatt, MWp = megawatt peak, PV = photovoltaic, SIP = solar irrigation pump.

Source: Asian Development Bank consultants estimate.

8.2.2   Impact on the National Electricity Grid

Variable renewable energy (VRE) sources such as wind and solar are intermittent in nature, as their output is dependent on external conditions. Integration of high shares of VRE sources into the grid, usually above 30%, may adversely affect the power system operation due to the variability, intermittency, and fast ramping nature of such energy. The share of VRE connected to Bangladesh's national grid in March 2023 was very low at 1.5%. This road map proposes installing SIP systems that would amount to 1,000 MWp solar capacity by 2031; if all is connected to the national grid and operates at the same time, this would represent less than 3.9% of total current power generation capacity. In reality, this share would be even much lower since it is very unlikely that all grid-connected SIP systems in the country export to the grid the total of electricity they produce at a given moment.

The *Grid Reliability Study for Integration of Renewable Energy into the National Grid of Bangladesh,* published in September 2021 by the Power Division, modeled and assessed the effects of different levels of VRE integration into the current national grid. In particular, the steady-state load flow, steady-state voltage stability, and dynamic simulations (frequency and angular stability) were assessed to understand the response of the grid to the integration of 10%, 25%, and 50% VRE into the national grid. The study concluded that load flow, voltage stability, frequency response, and angular stability for 10% and 25% integration of VRE would not negatively affect the grid and the grid would remain stable. Since the road map proposes the integration of much lower shares of VRE, the conclusion is that the rollout of the proposed road map does not represent any significant risk for grid stability.

8.2.3 Implementation Phases

The road map is foreseen as being implemented in two phases:

1. **The dissemination phase** covers 2023–2026 and aims to introduce SIP systems in all suitable *upazilas.*
2. **The market uptake phase** covers 2027–2031 and aims to stabilize the market for SIP systems while reducing subsidies to them.

Table 26 shows the proposed targets for SIP system implementation in the dissemination and market uptake phases, the number of diesel pumps replaced, the number of SIP systems and PV capacity installed, irrigated area, farmers benefited, diesel fuel displaced, and avoided carbon dioxide emissions in the dissemination and market uptake phases are shown in Tables 27 and 28.

Table 26: Proposed Number of Solar Irrigation Pump Systems

Type of SIP System	Pump Rating (kW)	Number of Systems in Dissemination Phase (2023–2026)	Number of Systems in Market Uptake Phase (2027–2031)
Small size	< 8 kW	1,340/year = 7,200	2,160/year = 10,800
Medium size	> 8 kW	2,160/year = 10,800	3,240/year = 16,200
Total	**45,000**	**3,600/year = 18,000**	**5,400/year = 27,000**

kW = kilowatt.

Source: Asian Development Bank consultants' estimate.

Table 27: Solar Irrigation Pump Systems and Photovoltaic Capacity Installed in the Dissemination Phase, 2023–2026

Type of SIP System	Pump Rating (kW)	Number of Systems	PV Capacity (MWp)	Irrigated Area (ha)	Benefited Farmers
Small size	< 8 kW	7,200	90	40,000	140,.00
Medium size	> 8 kW	10,800	310	130,000	400,000
Total		**18,000**	**400**	**170,000**	**540,000**

ha = hectare, kW = kilowatt, MWp = megawatt peak.

Source: Asian Development Bank consultants' estimate.

Table 28: Solar Irrigation Pump Systems and Photovoltaic Capacity Installed in the Market Uptake Phase, 2027–2031

Type of SIP System	Pump Rating (kW)	Number of Systems	PV Capacity (MWp)	Irrigated Area (ha)	Benefited Farmers
Small size	< 8 kW	10,800	136.8	63,000	200,000
Medium size	> 8 kW	16,200	476.6	197,000	590,000
Total		**27,000**	**613.4**	**260,000**	**790,000**

ha = hectare, kW = kilowatt, MWp = megawatt peak

Source: Asian Development Bank consultants' estimate.

8.2.4   Cost of Implementation

The total cost of implementing the road map is estimated at $1.8 billion. This amount includes equipment cost, installation and testing, warranties, and also the materials and civil works required to install water storage and distribution infrastructure.[77] Costs for grid connection, including energy meters, cables, and security equipment, are not included in these estimations since they are considered responsibility of the beneficiaries (farmers or sponsors). These costs are expected to be covered by public financing, including a national dedicated SIP Fund, and loans and grants from development assistance, as well as private financing facilitated and guaranteed by the government (Table 29).

Table 29: Estimated Costs and Source of Financing for Road Map Rollout
($ million)

Financial Source	Dissemination Phase (2023–2026)	Market Uptake Phase (2027–2031)	Total
Dedicated national solar irrigation pump fund	100	150	250
Grants and loans via development assistance	350	450	800
Farmers and sponsors (equity and debt)	300	450	750
Total	**750**	**1,050**	**1,800**

Source: Asian Development Bank consultants' estimate.

8.3   Financing Options

Financing will be needed for equipment and water storage and distribution infrastructure to meet the road map targets. The costs of grid connection are not considered in the financing options of this road map; these costs, including energy meters, cables, and security equipment, should not be borne directly by the farmers.

Bangladesh's GDP averaged 7.5% during the financial years 2016–2019 but moderated to 5.2% in 2020 due to the coronavirus disease (COVID-19) pandemic. To become an upper-middle-income country by 2031, Bangladesh will need to increase public and private investments. The government's Eighth Five-Year Plan (2021–2025) requires $750 billion to achieve its targets.[78]

---

[77] An estimation of unit costs is presented in section 8.1.2 (Cost of an SIP System).
[78] ADB. 2021. *Bangladesh Country Partnership Strategy (2021–2025)*. Manila.

The financing of this road map will require both public and private financing, so it does not put additional fiscal pressure on Bangladesh's public accounts. National initiatives aimed at major paradigm shifts are given priority by international climate financiers such as the Green Climate Fund (GCF). A National Fund may also offer financing for SIPs. The SIP rollout can be considered part of Bangladesh's energy transition pathway, a solution that combines climate mitigation with fighting poverty and inequality, so it should be attractive to these financers. International financing will require that the road map is officially adopted by the Government of Bangladesh. Private financing meanwhile can provide direct project investment in the form of equity and debt, or both.

The implementation of this 8-year road map is not only intended to cause the least additional financial burden on Bangladesh's national accounts and reduce the government's diesel import bill, but also be farmer friendly. While the percentage of financing contributed by farmers may seem low, these levels cannot be increased further without risking the competitiveness of SIP systems against irrigation using diesel and grid electricity.

8.3.1   Debt and Equity

Private sector financing is available to farmers and private sponsors, who can secure such financing through various means including bank loans, equity investments, or a combination of both debt and equity.

Debt

Farmers and private sponsors can acquire debt in the form of loans from commercial banks to complete the financing of their projects. It is recommended that the government facilitates private low-rate loans with other preferential conditions such as sufficient repayment time (recommended longer than 10 years) and safeguards in case of difficulties in making repayments. These commercial bank loans can be linked to the award of public finance and be channeled through local financial institutions. Under such a scheme, the burden on the government to administer the public finance part is largely reduced. The participation of local financial institutions, such as agricultural and rural development banks, lowers the burden on farmers since applications for the loan and public finance is done altogether in one local institution. The process of bundling the subsidy with the financing can be implemented through a banking network playing the role of the subsidy-channeling agency. Equipment suppliers connect customers to banks and help them in submitting subsidy and loan applications. The Bangladesh Krishi Bank and Rajshahi Krishi Unnayan Bank currently provide low-interest loans to farmer groups for their acquisition of agricultural equipment. Loans could be backed by green bonds issued by the government and guaranteed by development cooperation funds, such as those offered by multilateral development banks like ADB, the Asian Infrastructure Investment Bank, Islamic Development Bank, and World Bank.

Equity

Private sponsors would be expected to put down equity or an equivalent as requisite to obtain financing from commercial banks. Equity or equivalent serves to demonstrate to private financers creditworthiness for receiving the remaining finance as debt. This road map suggests equity offered should be equivalent to the same amount received as the bank loan for the project.

8.3.2   Grant Component

The road map proposes that public sector financing for SIP systems be organized through two complementary channels: a dedicated public fund (SIP Fund) that is set up by the Government of Bangladesh and loans and grants via development assistance as well as corporate investors who see benefits investing in energy transition.

Dedicated SIP Fund

The SIP Fund would be a public fund established by the Government exclusively dedicated to financing SIP systems to replace diesel pumps, becoming operational as soon as the road map is approved by the government. The fund can start with an initial minimum amount of $250 million. Having such a dedicated fund will enable the raising of further funds from other development partners and corporate investors.

Loans and Grants from Development Assistance

As the SIP Fund will not be sufficient to provide all the expected required public finance, the gap should be filled with loans and grants from multilateral development banks like ADB, Asian Infrastructure Investment Bank, Islamic Development Bank, and World Bank, along with corporate investors. It is estimated that this road map will need approximately $800 million in loans and grants through development assistance to complement the SIP Fund. The government could also consider facilitating special conditions (lower interest rates and longer tenure) for commercial bank loans that are dedicated to the installation of targeted solar PV power plants on agricultural land.

This road map is fully aligned with the implementation of Bangladesh's NDC; it aims to reduce the country's GHG emissions and prepare the agriculture sector for adapting to unavoidable impacts from climate change. Consequently, all official development assistance requested by the government for the implementation of this road map should be labeled as assistance for the successful implementation of Bangladesh's NDC.

8.4 Monitoring Implementation

The successful rollout of this road map requires that a mechanism is established for

(i)   the integrated assessment of projects;
(ii) financing a higher buyback price for power export to the grid from SIPs and grid interconnections for poor farmers;
(iii) monitoring SIP system execution and performance, and reporting and verification of the progress; and
(iv) measuring the effectiveness of the different supporting programs.

8.4.1 Integrated Assessment of Projects

All projects require a complex investment appraisal prior to financial approval. The installation of SIP systems are projects that have high up-front costs, with distributed revenues over long terms (>10 years), and require low-rate loans. A large variation of equipment is available in the market, in terms of both cost and quality. Good technical standards and mechanisms should be established to ensure that only high-quality equipment is used.

The application process linked to public financing and private sector loans should be as simple as possible and should avoid excessive bureaucracy. However, an integrated design and assessment of projects involves not only market, economic, and financial assessments but also thorough technical, environmental, and legal assessments that look into issues like types of crops targeted, sustainable use of water resources, arsenic levels of the water source, preparedness, and legal status of the sponsor (Figure 28).

Figure 28: Dimensions to Consider in an Integrated Assessment of Projects

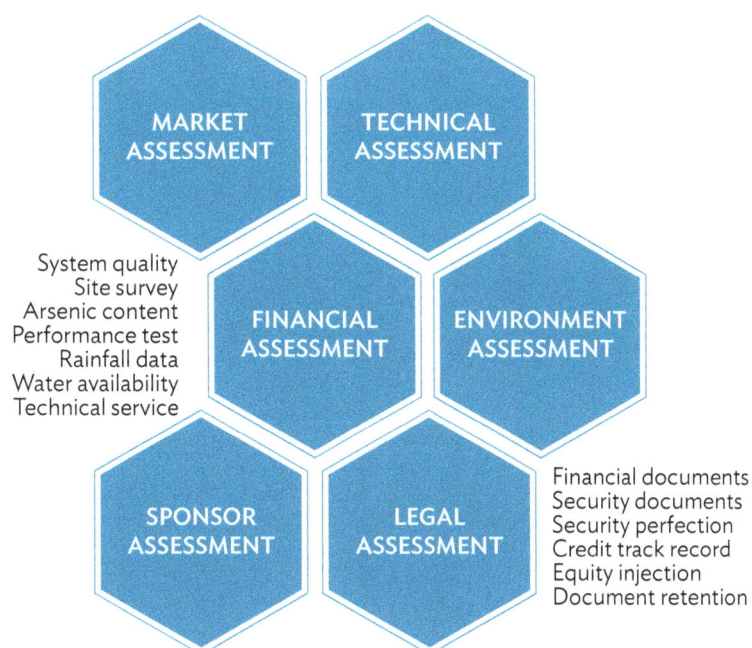

System quality
Site survey
Arsenic content
Performance test
Rainfall data
Water availability
Technical service

MARKET ASSESSMENT

TECHNICAL ASSESSMENT

FINANCIAL ASSESSMENT

ENVIRONMENT ASSESSMENT

SPONSOR ASSESSMENT

LEGAL ASSESSMENT

Financial documents
Security documents
Security perfection
Credit track record
Equity injection
Document retention

Source: Asian Development Bank consultants.

Provisions for after-sales service need also to be established, with local technical support and availability of spare parts in rural areas. Blocked pump foot valves and broken modules are the most common technical issues and require fast attention to avoid long service disruption.

A detailed list of issues to be assessed prior to financial approval is presented in Table 30.

Table 30: Assessment Activities to Be Carried Out prior to Project Approval

Assessment	Issues to Be Assessed
Sponsor	• In-house technical capacity for implementing and operating similar projects • Prior experience in implementing similar projects • Ability to inject minimum required equity and willingness to pay back private loan • Ability to provide collateral against private loan
Technical	• Verification of technical data and adequacy of equipment package to supply declared water output • History of disasters in the project area • Adequacy of site selection and arsenic check of water • Test of salinity and other toxic minerals • Compliance of equipment with technical standards that comply with standards set by Sustainable Renewable Energy Development Authority • Verification of equipment test certificates of equipment supplied or procured • Evaluation of price quotations and supplier selection

Assessment	Issues to Be Assessed
Market	• Demand assessment, including an assessment of the amount of land to be covered under the project • Farmers' preferences and interest in installing solar irrigation pump systems • Existing mode of irrigation (diesel/electric motor) • Existing major cropping patterns and number of crops per year • Existing irrigation expenses under each cropping pattern • Farmer composition (owner/hire)
Financial	• Evaluation of credit risk management • Verification of the minimum debt service coverage ratio • Project internal rate of return: 11%–13% • Equity internal rate of return: 13%–15% • Payback period > 10 years
Legal	• Registration and incorporation documents verification • Eligibility check of the borrower to conduct the business • Authentic identification document verification of the sponsors • Other compliance checks according to regulatory requirements

Source: Infrastructure Development Company Limited (IDCOL).

8.4.2 Effective Monitoring–Reporting–Verification

The monitoring, reporting, and verification (MRV) mechanism keeps the government informed about how the implementing programs are contributing to achieving the national policies (Figure 29).

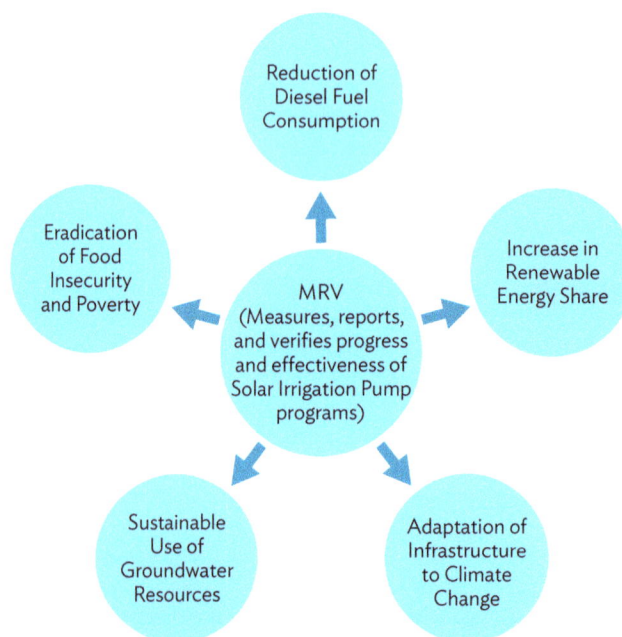

Figure 29: The Road Map Monitoring–Reporting–Verification Mechanism

Source: Asian Development Bank consultants.

The main issues to be monitored when project implementation starts are shown in Table 31.

Table 31: Assessments Activities to Be Carried Out during Project Implementation

Assessment	Issues to Be Assessed
Financial	• Execution of financial documents • Equity contribution • Perfection of securities • Submission of related documents
Technical	• Water well construction • Water distribution line layout and construction • Civil construction • Photovoltaic installation • Pump lowering and commissioning • Test to check water discharge performance • Other physical equipment checked according to the quotation

Source: Infrastructure Development Company Limited (IDCOL).

The benefits of establishing an MRV mechanism for this road map are as follows:

• clarity on what information needs to be developed and processed by implementing programs for their use by the different government agencies;
• more transparency in the management and reporting of information related to international financial support received;
• less likely double counting of SIP systems installed, diesel avoided, and emissions saved;
• makes comparable the results obtained by the different implementing programs;
• helps to identify technical and financial needs in implementing programs, as well as deficiencies and best practices; and
• helps to prioritize efforts during the remaining time to implement program.

The road map's MRV mechanism requires centralized coordination and involves all institutions implementing road map programs. It requires that implementing programs engage experts in energy and agricultural irrigation. The MRV mechanism should not be a burden on implementing programs but should instead help to timely deploy projects and encourage further investments.

The MRV mechanism should include measures to collect data that will be expressed in indicators the government may require. This information will typically include the number of SIP systems installed, their locations, the number of farmers and hectares benefited, the number of diesel pumps replaced, diesel fuel avoided, and GHG emissions saved. It will then be possible to compile the information in reports and inventories subsequently used by the government for review and analysis. These reports should have the level of detail required by national and international organizations funding the programs.

Indicators established by the MRV mechanism should be specific, meaningful, measurable, and cost-effective to harvest (if not already being collected). They should measure the progress and results of the implementing programs (e.g., the number of SIP systems installed or the capacity of PV panels installed), the outcome produced by those results (e.g., the number of farmers benefited, the number of hectares irrigated, the annual electricity exported to the grid), and the impacts these results have in the country (e.g., changes in national GHG emissions, amount of food produced, average income of farmers).

An adequate MRV mechanism presents an opportunity for Bangladesh to leverage climate financing for the further replacement of diesel pumps with SIP systems. It also helps ensure that support in policies and financing are effective.

9. Recommendations to Enable the Rollout

9.1 Key Findings

There is a unanimous view among government, nongovernment, and private sector personnel that the SIP rollout is necessary for Bangladesh, and the rollout is a cross-sectoral initiative that will require synergy among all key stakeholders to be successful. There is also consensus that attention must be given to lessen pressure on Bangladesh's groundwater: water-use efficiency must be increased and cropping patterns rationalized, considering water availability and sustainability of aquifers.

Although Bangladesh's cropped area has been reasonably stable, cropping intensity is gradually increasing, putting pressure on irrigation requirements. In 2021, more than 5 million ha were irrigated, with 72.65% irrigated during Rabi season (October–March) with groundwater using tube wells and the remainder irrigated with surface water using LLPs.[79] Measures that will help to achieve a more efficient use of water for irrigation include replacing diesel LLPs and tube wells pumps with SIP systems.[80]

9.1.1 Identified Success Factors

Pilot interventions investigating the feasible potential of solar irrigation have taken place in Bangladesh in recent years, overseen by the BADC, BMDA, BREB, BWDB, DAE, IDCOL, and RDA. Some success factors identified in current and past irrigation programs using SIP are the following:

- The design of an SIP system should be based on peak water demand, including crop water requirements and land preparation.
- Underground (buried) irrigation water distribution systems should be provisioned, giving due consideration to technical, social, and cultural practices of the water users.
- Prepaid metering systems with individual smart cards for water users in community/agency schemes should be provisioned to motivate the farmers to use the water economically, which helps to preserve national water resources.
- Grid-integrated SIP systems are very attractive to agencies and individuals because they provide additional income and reduce operational costs. Where net metering is impossible, farmers with SIP systems should be supported with grid interconnection in order to export their unused energy to the grid. Clear buyback tariffs should be established by the government to incentivize SIP adoption.
- Post-rollout support for O&M is currently lacking, despite solar irrigation projects having been executed for many years. Ministries leading the SIP rollout must find ways to provide post-rollout support, especially for the individual model. The involvement of the LGED should be considered.

---

[79] Includes shallow tube wells and deep tube wells.
[80] Pumps refer to centrifugal pumps (suction) or submersible (forced) pumps.

9.1.2   Suitable Areas for SIP Rollout

It is unlikely that the cropping area will increase in Bangladesh. However, crop evapotranspiration may increase due to an increase in cropping intensity through the introduction of short-duration crops and climate change. The increase due to climate change is estimated at 3%.

There is considerable scope for the SIP rollout to replace surface water LLPs. Existing sites where LLPs are used to lift water from a river or a tributary and the location of 64 rubber dams offer potential sites for SIP rollout. This road map identifies 75 *upazilas* as priority locations for surface water irrigation in which existing electrified LLPs can be replaced with SIP systems.

The rate of change in groundwater levels across all *upazilas* was analyzed using 20-year data of gravity recovery and climate experiment anomalies. This approach was preferred because publicly available groundwater-related government ordinances, reports, and journal papers on changes in groundwater levels across Bangladesh do not sufficiently quantify the rate of change in groundwater level to determine the statistical significance of changes. Total water storage of almost all (432 out of 495) *upazilas* recovered during the monsoon. Some cautionary signs indicate that terrestrial water storage (TWS) steadily declined in 63 *upazilas* and groundwater storage (GWS) steadily declined in 37 *upazilas*.

Out of the 495 *upazilas* analyzed, an MCA was carried out on 350 to prioritize them for SIP rollout. The other 145 were excluded from the MCA and are not recommended for SIP rollout since the government has not provided distance guidance for tube wells for them. These *upazilas* are mostly along the coast, subject to tides or floods, and have groundwater quality that is unsuitable for irrigation. The MCA was based on seven biophysical factors: groundwater salinity, arsenic, flood depth, tidal surge, tube well spacing, TWS change, GWS change, and crop water requirement. Based on the results, the 350 *upazilas* were prioritized into three priority groups:

- 158 *upazilas* as suitable locations for groundwater irrigation, where the current recharge mechanisms are adequate to sustain the rollout of SIP systems;
- 137 *upazilas* where water conservation measures are needed to support groundwater irrigation with SIP systems; and
- 55 *upazilas* where caution is necessary for the rollout of SIP systems.

The road map has excluded those 145 *upazilas* for which the government has not provided tube well distance guidance. In general, it is considered that salinity, flood, and tidal conditions in those *upazilas* may threaten the viability of SIP hardware.

It is noted that Bangladesh's water–food–energy nexus challenges are unique. Unlike countries where food security is affected by a lack of access to water, Bangladesh protects itself from water disasters and takes measures to minimize nonbeneficial water consumption. Bangladesh also has steps toward integrating the water-food-energy nexus by adopting solar irrigation and prepaid metering systems.

9.2   Institutional Arrangements

9.2.1   Regulatory Bodies

The MOPEMR and BERC have an essential role for enabling the rollout of this road map. As responsible bodies for setting the regulatory framework of the energy sector, these two institutions are in charge of assessing and proposing regulatory changes such as a net metering scheme suitable for SIP systems and deciding on the convenience and timing for increasing bulk tariffs for the purchase of surplus electricity exported to the grid by SIP system owners. These reforms would benefit from technical assistance provided by multilateral development banks to Bangladesh in the form of results-based lending or policy-based lending.

9.2.2   The Sustainable and Renewable Energy Development Authority

The Sustainable and Renewable Energy Development Authority (SREDA) is the nodal institution for the identification, promotion, facilitation, and overall coordination of all national renewable energy and energy conservation programs. SREDA's official mission is to

- coordinate and facilitate the development of renewable energy and energy efficiency,
- increase the share of renewable energy in the energy mix for reducing dependency on fossil fuel,
- take appropriate measures for energy saving, and
- assess new potential sustainable energy solutions.

To facilitate the implementation of this road map and align with SREDA's mission, it is suggested that the government enact legislation assigning SREDA with the responsibility of overseeing and implementing the road map. Moreover, SREDA should take on the following:

- The role of coordinating the efforts of various implementing agencies, such as BADC, BMDA, BWDB, DAE, IDCOL, LGED, RDA, distribution utilities,[81] to ensure the alignment of activities related to solar irrigation projects that receive public funding.
- The design and implementation of an MRV mechanism for this road map. SREDA should ensure its monitoring authority over the different SIP system programs by receiving reports from program implementing agencies and financial institutions providing support to farmers and sponsors. This will require the establishment of standardized reporting procedures and templates. SREDA is also responsible for confirming the design of the MRV mechanism and the MRV plan to meet the road map's requirements.
- Establishing an independent Technical Standard Committee for safeguarding the quality of SIP systems. The committee will be responsible for defining the technical specifications for all equipment financed under the road map, as well as the technical requirements for suppliers and installers. This committee can be established within SREDA, building upon IDCOL's experience.
- Providing procurement guidelines, with clear and adequate provisions for quality equipment performance, based on the technical specifications issued by the Technical Standard Committee. These provisions should be followed by purchasing entities, both public or private.

9.2.3   Implementing Agencies

Participating agencies, such as the BADC, BMDA, BREB and other distribution utilities, BWDB, DAE, IDCOL, LGED, and RDA, make sure they have strong technical provisions to guarantee proper O&M for all installed SIP systems, as well as the technical capabilities and appropriate organizational structure to meet their targets for SIP systems. BWDB should be included as a stakeholder in the implementation of all SIP systems; in particular, it should be involved in the selection of surface water sources and groundwater aquifers for each SIP system. In relation to the grid integration of SIP systems, distribution utilities must remain fully in charge of assessing and deciding the adequacy of grid connection on a case-by-case basis.

---

[81]   The distributing utility will be responsible for grid integration of future SIP investment projects in their respective franchise areas.

At present, BADC, BMDA, BREB, BWDB, LGED, and IDCOL are the agencies with relevant experience deploying SIP systems across Bangladesh. These agencies have expertise in different districts and have worked with surface and groundwater irrigation. In order to ensure effective oversight and compliance, the SIP road map calls for a level of decentralization in program management. This can be achieved through the local offices of experienced agencies or by assigning designated project developers who will operate within a defined set of framework conditions. It is essential that this decentralization in program management adheres to certain requirements, such as ensuring that all applicant projects have valid operating license issued by the corresponding *Upazila* Irrigation Committee and that there is no overlap in their service areas. No grant should be awarded to projects not fulfilling this requirement.

It is also recommended that the agencies concerned should enhance their coordination with the DAE to strengthen efforts in providing valuable agricultural extension services to the beneficiaries, especially for very small and small farms. To maintain highly productive and knowledge-intensive agriculture, agricultural extension is essential. Training in integrated pest and disease management, integrated farming systems, and technologies for conserving water and soil should be prioritized.

The road map identifies in Table 16 (section 6.6) potential lead agencies for some districts to lead the SIP rollout. The road map also identifies additional support to be provided for the lead agencies by other agencies. However, these recommendations need to be further assessed by these agencies, and institutional support needs to be confirmed or refined.

9.2.4   Financial Institutions

Mechanisms for financing the implementation of this road map will be needed, including the creation of affordable financial instruments by the banking sector and microcredit institutions. To finance their investment, project sponsors, farmer organizations, and individual farmers will need to obtain loans. There is a need for adequate lending products with low interest rates to be offered by the financial institutions. Standardization of such products is recommended.

9.3   Road Map Financing

9.3.1   Preferential Taxation to Lower Equipment Cost

Conducive policies supporting the cost reduction of equipment will be needed. Preferential taxation, tax reductions or exemptions, or accelerated depreciation are measures to mitigate the effects of high capital costs of renewable energy equipment in general, including SIP systems. These provisions could specifically include

- increasing incentives for exporting PV panels manufactured in Bangladesh and extending these incentives to also cover the production for the local market;
- keeping the tax waiver for imports of solar raw materials for the next 10 years;
- reducing or exempting the custom duty and value-added tax on pumps; and
- reducing the advanced income and trade taxes on pumps and solar panels.

It should be the responsibility of the Bangladesh National Board of Revenues, in collaboration with the Power Division and SREDA, to evaluate the viability of these tax benefits and create workable implementation strategies. These measures should be backed by a robust national renewable energy master plan to promote private sector participation.

Other provisions could include zero import taxes on renewable energy equipment for the next 5–6 years or longer, and tax holidays for 5–6 years to manufacturers who want to establish a base in Bangladesh that will create employment. Local manufacturers can be ring-fenced for government-sponsored projects so that the overall renewable energy sector development is not hampered. The International Renewable Energy Agency estimates that 18 job-years can be created per megawatt (MW) during a renewable energy project's installation and 0.3 job-years per MW during O&M. This means that for a combined 3,000 MW SIP system, ground-mounted solar PV system, and floating solar PV system, 54,000 jobs would be generated during installation and 900 full-time jobs for 20–25 years during the O&M of the plants. If Bangladesh concludes that floating solar PV systems are also a viable option for the country, then instead of a combined 3,000 MW program, a 10,000 MW program spread over a 10-year period will result in generating 180,000 jobs during installation and 3,000 permanent jobs during O&M. Tax incentives for renewable energy technology will enable private sector investments in other renewables, including battery storage, thus improving employment opportunities as well.

An investment tax allowance for the purchase of green technology equipment and income tax exemption on green technology systems and services could be explored as well. These green technology tax incentives must be designed to help drive the growth of Bangladesh's green economy by encouraging green procurement and the development of Bangladesh's systems and service providers' sector. They could include the following:

- Companies undertaking green technology projects for business purposes can benefit from a green investment tax allowance;
- Companies can take advantage of a green investment tax allowance when acquiring capital assets such as green technology equipment and systems; and
- Green income tax exemption, for companies undertaking new green technology activities approved by the government through SREDA.

9.3.2   Financing Paradigm Shift

Grid-integrated SIPs need to be promoted as this allows energy export during non-irrigation periods and reduces the financial burden of farmers. Empowering farmers to have a new source of regular income through selling renewable electricity to the national grid, while reducing their diesel dependence on irrigation activities and promoting a more sustainable and efficient use of groundwater resources, will need to be coupled with an international financing approach. Financing this paradigm shift requires a combination of dedicated government fund and financing preferably from the GCF, the Global Environment Fund, or via multilateral development banks. At the beginning of this road map, an initial fund of at least $250 million is needed to kick off implementation. This fund could receive government allocations from the avoided spending on diesel. Additional loans and grants from international development assistance will be needed for the overall targeted public financing. To obtain financing from international development assistance, the road map must be approved and adopted by the government for implementation, as this sends a clear signal that Bangladesh is on the path of energy transition.

9.4   Ownership by Farmers and Enhancing the Role of Women

Awareness of SIP benefits at all levels of government and among farmers is currently low. Therefore, targeted awareness campaigns are recommended. Farmers also need to understand insurance against risks of damage to SIP or theft of their solar panels. Raising awareness among farmers about the benefits of SIP systems and providing training and capacity building to government agencies and private sector can help promote the acceptance of new technologies in agriculture. In rolling out the SIP program, the government must make available funding and obtain commitment from all relevant stakeholders (farmers, agencies, and equipment suppliers) to ensure success of the SIP program. A deliberate effort is needed to persuade farmers to make the switch, even though farmer households show a great interest in solar systems for use in agriculture and beyond (such as for home lighting). This is a result of how these SIPs are marketed to farmers as well as the systems' cost. Seldom can farmers independently purchase SIP systems from shopping centers as an off-the-shelf item. Instead, access to the majority of SIP systems is restricted to those who take part in particular, frequently government-initiated programs. This points to the importance of addressing technology awareness and knowledge.

Awareness and capacity building are essential requirements for the success of this road map. It helps farmers, project developers, equipment suppliers, and financial institutions to understand and address the transition toward more sustainable irrigation. A sustained effort to educate farmers and boost public awareness about the benefits of transitioning from diesel pumps to SIP systems for irrigation is needed, particularly during the first years of road map implementation.

Awareness encourages changes in stakeholders' attitudes and behavior and helps them adapt to transition. Successful awareness-raising through television, radio, and newspaper will enable informed decision-making, play an essential role in increasing capacities of farmer communities, and empower farmers to adapt to change. Furthermore, conducting selected briefings in rural areas by bringing farmers who are already using SIPs to share their experiences will help farmers to decide to shift to SIP systems.

Encouraging the participation of women in SIP projects is important, especially in regions such as Dinajpur, Thakurgaon, and Naogaon. Special attention should be given to promoting women's ownership of SIPs and their involvement in providing clean drinking water. Plans should be developed to support and empower women in these areas.

9.5   Technical Support for Better Operation and Maintenance

9.5.1   Capacity Building

Capacity building for each investment project, in terms of enhancing skills for O&M, is essential for the sustainability of scaling up SIP systems. A key factor in determining success of a solar irrigation program is having skilled technicians at the local level. Implementing agencies will be required to provide strong technical support for O&M within their programs. This would be crucial as SIP system implementation is expanded throughout an area, and it would become even more crucial as systems age and require additional servicing and repairs.

SIP systems might be expanded but underperform or fall into disrepair, misuse, or disuse if there is no qualified local workforce. Unfortunately, when technically complex products are introduced in remote areas, this kind of thing usually happens. Only in larger cities can one acquire the necessary training and credentials to fix SIP systems without compromising manufacturer warranties; additionally, once trained, technicians might not be able or willing to move to rural areas where the market is much smaller.

The primary objective of required capacity building is to create a skilled workforce that will oversee O&M of SIP installations. An adequate capacity building program should aim to

- standardizing maintenance practices at national level;
- creating and distributing a variety of distance learning and classroom educational materials for various audiences, including technicians, master trainers, project developers, engineers, and policymakers;

- establishing uniform training programs through a network of accredited training facilities throughout Bangladesh; and
- creating a network of centers for entrepreneurship, technical training, and research and innovation to share best practices and encourage the transfer of knowledge.

In addition, beneficiary farmers need to improve productivity and income. Knowledge and skills gaps of farmers were identified in Bangladesh's Eighth Five-Year Plan as the main reason for low crop yields and low income from products trading. Increasing the productivity and profitability of crop production requires training. However, inadequate training is widespread, even in Bangladesh's most agriculturally active districts. Training should be arranged together with DAE based on the necessity of the farmers. DAE activities can add value to the process and help sustain food security. The most beneficial trainings to farmers include the following:

- Field crop production: Weed management, cropping systems, water management (alternate wetting and drying), integrated farming, seed production, crop diversification, and nursery management.
- Crop protection: Integrated pest management, integrated disease management, biocontrol of pests and diseases, and production of biocontrol agents and biopesticides.
- Soil health and fertility management: Soil fertility management, soil and water conservation, management of problematic soils, soil and water testing, integrated nutrient management, and fertilizer application.
- Vegetable production: Production of low-volume and high-value crops, production of offseason vegetables, exotic vegetables production, seedling raising, export potential vegetables, vegetables grading and standardization, protective cultivation (greenhouse, shade house), training, and pruning.
- Plantation crops: Production and management technology, and processing and value addition.
- Tuber crops: Production and management technology, and processing and value addition.
- On-farm production of inputs: Bioagents production, biopesticides production, biofertilizer production, and vermicompost production.
- Agricultural engineering: Repair and maintenance of farm machinery and implements, and postharvest technology of vegetables and fruits.
- Marketing and transportation.[82]

Finally, maintenance, operation, and management costs should be recovered from irrigators. Prepaid meters based on volumetric water supplied are effective for the cost recovery of community/agency-based systems.

9.5.2 *Technical Standards*

The road map recommends that the government prepares clear technical standards that are applicable nationwide in the purchasing and installation of SIP systems. The observance of technical standards should be a mandatory requisite for obtaining any type of public support. Failure to comply with these technical standards should lead to the immediate termination of government support.

Currently, IDCOL is the only financial institution that has developed such standards. Instead of reinventing the standards, SREDA should ensure that these standards are adopted by other agencies and financial institutions in scaling up the SIP program. IDCOL standards include technical specifications for pumps and motors, PV panels, controllers, underground pipelines and header tanks, pump houses, and water wells.

---

[82] M.S. Rahman et al. 2018. Assessment of Training Needs on Crop Production for Farmers in Some Selected Areas of Bangladesh. *Bangladesh Journal of Agricultural Research.* 43(4). pp. 669-690.

For all these elements, the standards specify technical requirements and test guidelines. These standards also establish the warranty provisions that suppliers and installers must provide. These technical standards should be updated and improved on a regular basis as new technologies are developed and emerged in the market. These should be accompanied by technical specifications for grid interconnection infrastructure, such as cables, fuses, switchgears, transformers, and protection schemes for both low- and medium-voltage interconnections; and quality of exported power at the point of interconnection, including specifications for voltage fluctuations, flickering, frequencies, harmonics, and power factors.

9.6   Lowering the Cost Burden

The general economic principle for the success of the road map rollout is that the SIP-based irrigation cost of targeted installations should not exceed the cost of farmers' existing diesel-operated pumps. To lower the cost burden, two recommendations related to the compensation of exported electricity and safeguards for farmers in times of financial distress are given.

9.6.1   Compensation Mechanisms for Exported Electricity

The implementation of the proposed road map requires clear provisions for the compensation of renewable electricity exported to the grid. Such provisions should result in suitable schemes offering a guarantee for the purchase of electricity by utilities or net metering at reasonable rates.

SIP systems replacing diesel pumps can export their surplus power to the grid when pumps are idle, in particular during the off-agriculture season. The exportable electricity to the grid by the SIP systems proposed in this road map is estimated in 480 GWh per year.

The conditions for the purchase of this electricity by distribution utilities are defined under the *Guidelines for the Grid Integration of Solar Irrigation Pumps 2020*. Any owner of an SIP system or a developer with legal permission from the owner can apply for grid integration to the local distribution utility to sell the surplus of electricity produced by the SIP system up to 10 MWp. In case of grid integration, rules and standards prescribed by the concerned utility must be followed. SIP capacities for connections at 11 kV cannot exceed 70% of the transformer's capacity, and if it does, the transformer capacity will need to be upgraded. The distribution utility will ensure uniform system connectivity of the feeder substation to three-phase connections. Applications for single-phase connections at low voltage are also accepted. The distribution utility will pay the exported electricity at the 33 kV bulk electricity tariffs approved by BERC (Table 18) or at a higher rate as stipulated by BERC following discussions between BERC, the Power Division, and the Ministry of Finance. The applicant will be responsible for paying the cost of the meter and any additional fees. Applicants will also be requested to pay for any change in the security management of the distribution utility such as for fuses, switching gears, or transformers.

Grid-connected SIP systems are allowed to import up to 1 kWh of electricity per kW AC from the grid to keep the system components operational (i.e., inverters, energy meters, relays, magnetic contactor, and lights). Under current regulations, connected SIP systems are not allowed to import additional electricity from grid for other uses and will not be able to apply for a new power connection to feed the same premises.

It is recommended that the *Guidelines for the Grid Integration of Solar Irrigation Pumps 2020* be revised as soon as possible. The issue is that all SIP systems will have a reasonable out-of-order time during the year (due to equipment failure, breakdown, regular maintenance, prolonged poor weather, and so on) and farmers need to have a backup alternative. If farmers cannot import electricity from the grid to operate their pumps in those situations through a net metering scheme, they will be obliged to use diesel pumps as backup. This defeats the purpose of having SIP systems. However, in the future, there should be regulatory provisions for these SIP systems to use grid electricity during non-sunshine hours (cloudy days and evenings). After 2030, it will be desirable for all irrigation pumps to be connected to the main grid and be able to operate in hybrid mode, with solar as the priority. The guidelines should be revised to allow import of electricity from midnight to 7:00 a.m. or any other off-peak times of the day or year. This will also reduce dependency on standby diesel pumps by individual farmers.

It is also recommended that a mechanism to subsidize the grid connection cost of existing (already installed) and new solar pumps to smaller and poorer farmers is established by the government, since these farmers will not be able to afford the extra financial burden of paying for the grid connection of their SIP systems. This mechanism should cover the costs of connection (other than meters), ancillary services, and security management of the distribution grid. The government should make this mechanism clear to all distribution utilities participating in the SIP program. This mechanism could include a direct subsidy covering part of the connection cost combined with a financing scheme against the benefits of future exported electricity. Such a financing scheme could be offered to farmers by distribution utilities via a price discount in the electricity they purchase.

9.6.2   *Safeguards for Farmers in Times of Financial Difficulty*

Beneficiaries of this road map are small and very small farmers, usually with low income and with few or no savings. This is a barrier for these farmers in paying any equity or making any relevant down payment. These farmers usually have low creditworthiness to get loans since their situation makes it difficult if not impossible for them to pay back loans in times of poor harvest.

Policies should be put in place to assist farmers in repaying loans, especially during lean harvest or after disasters like floods and cyclones. Insurance for agricultural products or loan repayments are two examples of such measures that the SIP Fund may help with.

9.7   Energy Storage for Grid-Connected Pump Systems

Battery energy storage systems could be offered in the second phase of this road map (i.e., after 2031) as optional equipment for grid-connected, pure solar systems and for dual solar-electric pump systems. Installing energy storage could delay investments in new peak power generation capacity or the modernization of transmission and distribution electricity grids while increasing the availability of ancillary services within the grid. For instance, arbitrage opportunities may be facilitated by energy storage. Arbitrage is the practice of storing energy during times of low demand and low price so that it can be sold to the grid at times of high demand (peak hours) and higher price.

Ancillary services refer to the support services that are necessary to maintain the stability, reliability, and quality of electric power systems. These services are provided by the power system operators to ensure that the power grid operates efficiently and effectively. Ancillary services require quick response times, but they should only be offered momentarily. Regulation for energy storage and for the provision of ancillary services, which are important for security, will be needed. This regulation should include clear technical codes and compensation mechanisms for the following key ancillary services:

- frequency regulation, required to balance variations in the supply and demand of electricity;
- load following, which is comparable to frequency regulation but extends over a longer time frame—for example, from 15 minutes to 24 hours;
- voltage support, which ensures voltage levels in the distribution and transmission networks;
- black start capability, which is required to reactivate power plants following a system failure;
- spinning reserve, which can be accessed online in less than 10 minutes and used to compensate for unanticipated variations in supply or demand; and
- non-spinning reserve, an additional offline reserve capacity that can be instantaneously activated and sustained for several hours.

For SIPs to provide ancillary services, it is important that BERC establishes clear rules and regulations. To attract private sector investments in these systems, fees should be paid if the ancillary services are provided during peak periods.

www.ingramcontent.com/pod-product-compliance
Lightning Source LLC
Chambersburg PA
CBHW042033220326
41599CB00045BA/7285